D0645971

JUNGLE FLIGHT

Spiritual Adventures at the Ends of the Earth

Stories from
JAARS

By Dane Skelton

Copyright © 2009 by Dane Skelton

Jungle Flight
Spiritual Adventures at the Ends of the Earth
by Dane Skelton

Printed in the United States of America

ISBN 978-1-60791-922-3

All rights reserved solely by the author. The author guarantees all contents are original and do not infringe upon the legal rights of any other person or work. No part of this book may be reproduced in any form without the permission of the author. The views expressed in this book are not necessarily those of the publisher.

Unless otherwise indicated, Bible quotations are taken from The Holy Bible, New International Version. Copyright © 1973 by International Bible Society.

Photo credits:
cover photo: Tim Harold
pp. 50, 51 Tim Harold
pp. 54, 64 Ben Greenhoe
p. 90 Dennis Bergstrazer
p. 72 Woody McLendon

www.xulonpress.com

FOREWORD

❧ ❧

J ungle Flight will ignite your imagination and inspire your heart
to read about survival and rescue missions made possible by the
miracle work of Jesus Christ, the Pilot of our souls. This collection
of short stories will allow you to peer into the cockpits of airplanes
set apart for missionary service in the Name of the Lord and learn
how aviation has contributed to carrying out the Great Commission.
As a pilot myself, I have been amazed at how God has used planes
and helicopters as effective tools to elevate Bible translation for the
purpose of evangelism in the most remote parts of the world.

Franklin Graham
President and CEO
Samaritan's Purse
Billy Graham Evangelistic Association

In Memory
Schafer J. Skelton Jr.
Dad, you would have loved JAARS.

INTRODUCTION

W e don't attach the word "adventurer" to the word "missionary" anymore, thinking, perhaps, that all of that happened long ago to men with names like Stanley and Livingstone, or even Nate Saint and Jim Elliott. But those two words are still connected in ways I would never have dreamed until I went on an adventure of my own, a journey taken to encourage my friend David Reeves and some of his fellow missionaries serving with JAARS in Southeast Asia. JAARS, formerly Jungle Aviation and Radio Service, was founded by the same brilliant innovator and evangelical entrepreneur who founded Summer Institute of Linguistics and Wycliffe Bible Translators, William Cameron Townsend. Townsend committed himself to developing an organization that could train and equip jungle aviators to serve the needs of translators in speed and safety. JAARS is the result.

So could I help? The idea was to take a gift – a tool to help service the airplanes – preach a little and generally try to be an encourager to some people working very hard to serve God in a difficult place. Little did I realize that the gift, in fact, would be given to me.

Once in Southeast Asia, the things I saw and the stories I heard had a profound impact on my spiritual life. I asked David, "Is anybody telling these stories? Is anybody writing them down?"

To which he replied, "A few are, here and there. But none of us really have time." The stories were compelling, dramatic, engaging, and sometimes miraculous. They needed to be told. With the grace of God I decided to tell them.

The Betsch family boards for their vacation flight.

1

Flying through the Eye of the Needle

❧ ❧

For he will command his angels concerning you to guard
you in all your ways; they will lift you up in their hands, so
that you will not strike your foot against a stone.
Psalm 91:11-12

Pilot Keith Betsch knew he was going to strike one of the trees in front of him. He just had to decide which one.

Betsch was in a bad situation for any pilot. But the danger before him was multiplied to infinity by the passengers on board his twin-engine Piper Aztec. His wife, their four children, and his parents were in the plane as Betsch desperately tried to fly up the mountain in his windshield. Attempting a go-around on this high altitude airstrip was the worst mistake he'd ever made as a pilot, made in the worst of all possible places, for no place is more demanding of a bush pilot's skill or his airplane's performance than the jungle-clad mountains of Southeast Asia. Betsch was certain that he was about to crash on one and hurt or even kill everyone he loved.

Growing Up with a Dream

Keith loved being a mission aviator. He had dreamed, planned, and worked to become one since he was a boy growing up with missionary parents in the 1960s, first in Vietnam, then in Indonesia.

He had known since sixth grade, when a missionary pilot spoke at his church, that this was God's call on his life. He couldn't imagine anything more fun or meaningful.

When he told his parents his dream, they encouraged him. They also told him it would be impossible for them to help pay for it. "But if God calls you to it," they said, "he will make a way for you."

So Keith had prayed, "Lord, if you want me to be a missionary pilot, you're going to have to provide." And provide he did, sometimes miraculously.

Betsch, not quite six feet tall, stocky, blonde, and with a boyish grin, had gone to LeTourneau College in the 1980s, earning his pilot's and mechanic's licenses there. One of the ways God provided for him was through the campus ambulance service. Betsch could work sixty to seventy hours a week and still have time to study. He didn't have much in the way of a social life, because he always had to study ahead in case a call came in the middle of the night before an exam or a paper was due. But it paid the bills. After Keith graduated he served as assistant chief flight instructor, another paying position that built flying time and also gave him management experience.

But God's financial provision wasn't only through paying jobs. In fact, Keith never knew where all the money came from. He remembered standing in line once to pay his tuition, knowing he had no funds in his school bursar's account, hearing his friends ask, "Keith, why are you standing here? You know you don't have the money to pay."

Keith replied in his quiet, unassuming manner, "Well, God has me here. We'll see what happens."

And he got up to the table, gave the registrar his name, the man consulted his notes, looked up and said, "Your bill has been paid."

Tears still came to Keith's eyes as he remembered numerous events like that over the years. God's provision had been so consistent that it got to be a running joke among Keith's friends: "Hey, if you ever need cash just go look in Betsch's mailbox. There's always some in there!"

By the time he finished at LeTourneau, Betsch had no doubt that God had called him into mission aviation.

Island Paradise

That calling led Keith to JAARS where, after rigorous training, qualifying, and partnership development, he was assigned in 1991 to a beautiful island in Southeast Asia. His experience at LeTourneau prepared him to serve as aviation program manager. His wife, Lois, whom he had met and married while still in Texas, joined him.

Keith enjoyed the flying. Most of it was island hopping in the Robertson STOL (Short Takeoff and Landing) equipped Piper Aztec, a workhorse of an airplane with twin two-hundred-fifty horsepower engines. At sea level, where most of their destinations lay, the Aztec could easily operate out of twenty-four hundred foot government owned airstrips. Many of those strips were built with crushed coral by the Japanese during World War II, some with bomb damage still evident.

Even though the airstrips were marginally long enough and at sea level, landing in these villages on the edge of the world still had peculiar challenges. Cows roamed all day on most of the airstrips, and the pilot had to run them off before landing. It became a routine: Make the initial pass at about five hundred feet to inspect the strip and confirm the presence of bovine obstructions. Fly out over the shallow blue green water and climb back up a couple hundred feet. Make a teardrop turn. Line up for the runway, wheels up, no flaps, push the nose over, and leave the throttles in. Roar down the strip twenty feet off the deck with all five hundred horses bellowing like a herd of bulls, and pull up to clear the trees at the end. Zoom, zoom, cows gone!

JAARS procedures require pilots, once they are committed to a landing, to hit a cow or any other animal in the way, crunching the airplane if necessary, rather than risk human lives with a low clearance, low speed, last second go-around that might end in a deadly stall. Keith never had to hit a cow.

Aviators get attached to their airplanes, and Keith loved the Aztec. But he was also attached to the location. He remembered flying out over the bay of his home island, seeing the beautiful coral sand on the shore extending into the surf, still visible many fathoms beneath the waves, peering down and down through the deep cobalt of the ocean, the whales playing and the reefs casting shadows in varying shades of blue, and dreamed of scuba diving and exploring those indigo depths. His home base was lovely too, a peaceful place, orderly, with decent people all around who had learned to thrive under various governments and with different religions in a land where the rhythms of fishing and tides and seaside life had gone on for millennia. But it wouldn't last.

Betsch woke one morning, two and a half years into his four-year assignment, with an excruciating pain in the left side of his face. He didn't think much of it at first, took some ibuprofen and expected it to go away. But the pain wouldn't fade. If anything it grew stronger until it became disabling. Was he suffering from multiple sclerosis? A brain tumor? There was no paralysis, no other symptoms. The mission doctor had one other guess, but if his guess proved to be true, it was a very rare diagnosis, one hard to pin down. The Betsches were forced to pack up, leaving their third-world "paradise" for better medical care back in the United States.

Tic Douloureux

When he finally made it to a doctor in the States, the news was worse than Keith had imagined: trigeminal neuralgia, Tic Douloureux – a degenerative disease of the facial nerve – extremely painful and incurable and exactly what the mission doctor had feared. Medications would hold it at bay for a while, but the final solution meant clipping the nerve that runs from the back of the head to the face, leaving him with an awkward smile the rest of his life. Worse, from Keith's point of view, his flying days were over.

Surprisingly, the grief was milder than Keith had expected. Having dreamed and worked from the sixth grade till his early thirties to become a mission aviator, achieving it and then losing it, Keith thought that he would feel much lower than he actually did. But the calm assurance that God was in control, the assurance he had known at LeTourneau, was still with him.

Before they'd left their island home, Keith had arranged for a good friend and oral surgeon to meet with him as soon as he arrived Stateside to help him rule out any tooth or jaw problems that might be causing the pain in his face. The doctor scheduled appointments the day the Betsches arrived, Christmas Eve 1993, for Keith and one other patient, a middle-aged lady suffering similar symptoms. His friend's examinations confirmed the worst, and knowing that he could do nothing for them, he introduced them to each other and referred both to the same neurosurgeon.

Keith and his fellow sufferer with "Tic D," as they came to call it, continued to cross paths at the neurosurgeon's office from time to time, checking in with and encouraging each other. But while her course of treatment followed the normal path, Keith's doctor seemed more surprised with each visit.

One day in April of 1994, Keith pulled into the medical building parking lot feeling better than he had in a long time. He walked in, signed the registry, and sat down next to his favorite potted plant with a *Sports Illustrated* magazine. He didn't really like *Sports Illustrated*, but it was better than *People*. His fellow sufferer came in just as he was called back to see the doctor. They nodded to one another as Keith opened the door and followed the nurse to the examining room.

After a short wait, the doctor came in, went through his routine, asked a few questions, and then – with a strange look on his face – sat down across from Keith.

"This doesn't happen," he said. "There is obviously a greater physician than I working on you here. You might as well quit wasting your time coming to see me. You're healed."

Keith walked out into the waiting room as if in a dream, tears and gratitude welling up from inside, and saw his fellow sufferer. "How's it going?" he asked.

"I just had the nerve clipped," she said with an uncomfortable smile.

Keith felt a kind of reverential awe. *I'm no better than she is*, he thought. *That could have been my path.*

"And you?" she asked in return.

Keith didn't have the heart to tell her his news. "OK for now" was all he could manage. He slipped quietly out of the office, keeping his composure until he reached the privacy of his car. Then he put his head on the steering wheel, thanked God, and wept in both joy and sorrow.

During his illness Keith had been serving as the assistant to the aviation manager at the JAARS headquarters in Waxhaw, North Carolina. As his health improved, he and Lois began to ready themselves to go back to the field, eager to return to their island paradise. But it was not to be.

Flying through the Eye of the Needle

The door to paradise was closed. A replacement JAARS pilot, Paul Westlund, was now in the middle of his term there, and the Betsches were needed somewhere else, a more difficult place for flying *and* living. Keith's earlier assignment was mostly island-hopping; this one required more mountain flying. Whereas the town where they lived before was peaceful and happy, this was a place with less sophistication, greater tensions, and more social unrest. They had never needed security in paradise, but here they needed burglar bars on the windows and caution in the streets. It took a while for Keith and Lois to warm to the idea. But finally, in 1996, after eighteen months on the mend in Waxhaw, they headed east.

The mission was different on this island too. They used the Aztec a lot more for commercial flying, hauling supplies and passengers that helped the organization pay for translator flights. Takeoffs and landings at some of these airports were marginal in the Aztec. Mountain airstrips are by their nature higher than their seaside counterparts; the air is thinner, the lift over the wings less robust, the props less efficient, and air flow over the rudder needed to maintain minimum controllable flight less effective. Betsch had always enjoyed a challenging flying environment. But at many airstrips on this island a single mistake could be deadly. He was learning a new meaning for the word *caution*. But just as optical illusions can confuse a pilot's vision, distractions can confound his judgment, as Keith would soon discover.

Like many pilots' wives, Lois Betsch hated flying. But the stresses of living and working on that equatorial island had persuaded her that a little retreat would be worth it even if it meant a trip in the Aztec. Besides, Keith's parents were visiting. What better time to take their four children, two boys and two girls, and Grandma and Grandpa out to a beautiful, secluded mountain town for a little R&R? The air would be cleaner, the temperatures cooler, the scenery beautiful, and best of all, no phones would be ringing. The destination airport had once been a base for Mission Aviation Fellowship which meant the runway was plenty long, and there were a couple of well built, small houses to rest in. She could handle the one-hour flight, and their budget could take the cost of the trip. Lois was ready to go.

Keith agreed, but he was also aware of a minor property dispute in their destination town. The people who owned the land that included the airport were complaining that the government was not paying enough for use of the strip. *But that kind of stuff goes on all the time*, he reasoned within himself. *No reason to call it off.* Keith filed a flight plan, weighed the baggage, preflighted, and loaded the plane. Then he strapped his children in the back with Lois and Grandma, climbed in next to his dad who was in the right seat, and flew off toward the mountains.

Once airborne and on the way to their retreat, they watched the sky grow thick. Rain pelted the cabin and distorted the view through the windshield. Keith had checked the weather reports and knew it was clear over their destination, but the storm made Lois uncomfortable. "It's going to be OK, Honey," he reassured her through the intercom. "No problems." But the rain continued and the ceiling hovered around three thousand feet.

Finally the windscreen cleared. Subdued sunlight filtered through the still overcast sky, the Aztec hummed along, and the hardest part of the trip was behind them – or so they thought. But Keith couldn't raise anyone on the radio at their destination.

Strange, he thought, *but I'm sure it's nothing. Somebody will be there by the time we arrive.* The weather reports from surrounding areas continued to be good so he pressed on.

Someone on the ground heard the familiar drone of an airplane in the distance. Keith watched as he ran out onto the airstrip, hopefully to tell the people there that a plane was on its way. But he didn't head for the radio shack, apparently, because there was still no response to Keith's calls. Soon the plane was circling the little mountain town.

Keith thought, *Well, surely somebody will see me now and get on the radio.* But nobody responded. Nobody picked up the microphone. *That's odd,* he thought. *Something is not right.* Suspicion rose like a mist from the river of his subconscious mind and began obscuring his judgment.

Keith circled again and saw people on the airstrip. But they walked off, so he circled a few more times and then went down to about one thousand feet for a look.

The airstrip was plenty long, but it sloped up. Where the runway ended, the mountain continued up to fourteen thousand feet or thereabouts, and another ridgeline a few hundred meters to the right ran parallel to the airstrip. Its beginning marked the airborne committal point, about three hundred feet above ground at the end of the runway. That meant that once he passed a certain point in the approach pattern, even though he wouldn't yet be in the landing flare, he had to land. There was no "go-around" option after that.

So he thought, *OK, looks like we can land,* and reduced power to enter the landing pattern.

As his airplane slowed, Keith rested his thumb on the flat control handle at his right and pushed in the first notch of flaps. His practiced hand moved without thought now to the wheel-shaped lever to the right and, as he came abeam of the touchdown point and airspeed continued to fall, pushed down, lowering the gear into place. He heard and felt the familiar *thunk* as the lever came back to center and glanced at the three green lights indicating gear down and locked. He pulled back on the throttle levers and reduced power to thirteen inches of manifold pressure, continuing to slow and descend, feeding in more fuel with the mixture levers. He turned on the base leg and added another notch of flaps, and the Aztec reshaped her wings to grab greater swaths of air. Betsch's hand now moved to the blue prop control levers between the throttle and mixture levers, pushing them all the way in and now back to the throttles closest to him, his body fully alert, feeling for anomalies in the air currents, right hand ready to adjust power on approach. Recheck for three green lights, gear down and locked, then turn on final and back to the flap lever for the full aileron drooping effect of the Robertson STOL kit. He was now past the committal point, the point of no return. Just as he began to ease back on the throttles and pull back on the yoke for the flare, he looked up the airstrip and realized there was a huge pile, a wall of gravel, rocks as high as the nose of the airplane, right across the middle of the runway.

Everything seemed to slow down at that point though his mind jumped to warp speed. He knew that they could walk away from landing and hitting that wall, but it would destroy the airplane. He wasn't in the flare yet, so he thought, *I'll just touch the power, jump the wall, and set down on the other side and try to get this thing stopped before we run off the end of the strip.*

He eased the throttles back in to clear the obstacle when a hailstorm of thoughts rampaged through his mind: *Should I be landing here at all? Is it safe for my family with this dispute going on? Will we get caught up in the middle of it? What if I can't get the airplane out of here? It's not safe for my family here!* Instinct took over as Keith fire-walled the throttles for a go-around.

But there was no go-around option from here.

Betsch's right hand now raced over the throttle quadrant while his left gripped the yoke. Mixtures at seventy-five percent or the engines would choke at this altitude, take a notch out of the flaps, get the gear up to reduce drag, training and instinct took over as he coaxed every ounce of climb out of his ship. They were already at sixty-eighty hundred feet density altitude, the plane was fully loaded, and Keith knew it could not out-climb the mountain in his windshield. Lois knew it too. The mountain was all she could see.

The end of the runway loomed, but there was a shallow valley to his right, falling just below the tall ridge farther to the right, which blocked his path to freedom. He gave the Aztec a touch of aileron and slid his machine over the housetops and into the valley. She was flying on the edge of her envelope now, mushy, sluggish, and at sixty-five knots airspeed just barely above the stall. He fought the urge to yank on the stick, knowing a heavy hand on the controls would spin them in and kill them all. That's when he saw the trees. A huge one on the left, at eye level. His mind registered it as some kind of hardwood. A shorter tree loomed to the right, still at eye level, but softer, maybe like pine. He aimed for that one and heard the ***thump*** as the top three feet of it dented the right wing.

"Did we just hit a tree?" his son yelled. But Keith was too busy to answer. He knew he was out of options. The Aztec was about to run out of airspeed and altitude and hit the mountain all at the same time. The valley they were over was strewn with large rocks, but there was a soccer field next to a school building just beyond. It was his only choice. He would make a controlled crash down there. He'd destroy the plane, but his family might survive.

Keith had his hands on the gear lever when something caught his eye about three hundred meters to the right, an opening through the trees!

A gap on the ridge, a little over twice the width of his wings, beckoned to him. He made another gradual, oh so gradual, turn to the right, barely twenty feet above the ground now, and headed for the eye of the needle. His family looked out the windows as the soccer field and school building flew under their right wing, trees soared one hundred feet above them on the mountains to either side. Keith eased the nose up, trading his last five knots above stall for a few more feet to clear the ridge and then, like a hawk diving for her prey, the old bird threaded the needle and nosed over, zooming into open sky above the deep valley on the other side.

Everyone in the cabin began to breath again as Keith reconfigured the Aztec for a cruise climb to let the engines cool. Then he set a course for home.

Epilogue

About a week later, Keith and chief pilot Tom Beekman went back to the airstrip to "get back on the horse" and reinstate Betsch to flying status, standard procedure for a near miss. As it turned out, the land dispute was real, but it was not the reason for the rocks on the runway. The airstrip was undergoing maintenance, and the workers had failed to mark the threshold with a big white X to show that it was closed or to radio anyone about the work. Keith and Beekman did go-arounds from just prior to the abort point, and committed landings, where Tom would call for a go-around after the committal point to be sure Keith would continue to be committed to land. They checked the altitudes and distances and did the math. Under *ideal* conditions the Aztec was capable of making that climb. But Keith felt the same way that he did on the day he learned that he no longer had Tic D. It was the grace of God that got them home. The people never said why they did not answer the radio.

Keith came back to the States and the JAARS Center in 2005 to serve as an administrator. He is now the Senior Vice President of International Services, providing vital support for Bible translation work worldwide. In an early 2009 interview, Betsch, a professional pilot who could be making a comfortable living in the secular world, explained what keeps him in mission work.

"Jungle flying could cost you a lot, your life even. Bible translation is a cause that's worth that. It is also the most fun I've ever had. You can't wipe the grin off your face. Flying high performance airplanes into difficult spots that they alone can handle is a great thrill. Nothing else compares to it for me. But if that's the reason you want to do it, if the thrill is the only motivator, you won't stay. The cost is too high. Life in developing countries is too difficult.

"Even so, it was very tough to give up the flying and come back here to do the administrative stuff. But the bottom line is that the airplane is a tool to get the message of God's love and his Word in the mother tongues out there to the people who are still waiting. If I can have a broader impact for that doing administrative work, then this is where I'll be. But I would still like to be out on the field, flying the airplanes. When my kids get educated and if JAARS doesn't need me here in Waxhaw anymore, I'll go back."

A jungle driveway

2

Zaccheus of Mamaindé

❧ ❧

Submit yourselves, then, to God. Resist the devil, and he will flee from you. James 4:7

CB was the village thief. But Jon Caton didn't know that. Jon was in the Mamaindé region of western Brazil on a mission with JAARS Vernacular Media Services (VMS) to record *The JESUS Film* into the local dialect.

Wycliffe Bible Translators is known worldwide for putting the Bible into the heart language of Bibleless peoples. Every language group that Wycliffe serves will, by God's grace, one day have their own Scriptures. But their translators discovered something that, at first, seemed discouraging. Oral cultures (people groups with no written language that depend on storytelling for history and other important knowledge) often mistrust print. They won't believe a story until they hear it told by someone they trust or perhaps see it on screen. VMS bridges this gap. It puts the Scriptures into use in the culture immediately, via audio recordings, filmstrips, or video in a format that is more "user friendly" to oral cultures. Seeing or hearing the story builds trust and often stimulates a desire to learn to read.

The Mamaindé (pronounced mah-mine-DAY) were like this. The Mamaindé Gospel of Mark had been in print for over twenty years, translated by a Wycliffe couple who lived in the village from

1959 to 1979. Yet the Mamaindé never responded to the message it contained. In 1989 Wycliffe translators David and Julie Eberhard joined the Mamaindé to continue the work. By 1998 they still saw no meaningful response.

Enter Jon Caton and a VMS colleague from JAARS. Through an arrangement with *The JESUS Film* producer, Campus Crusade for Christ, VMS had permission to translate the film for smaller language groups like the Mamaindé.

The crew first flew into São Paulo, Brazil, and then to Cuiabá. From there, a twelve-hour bus ride took them to the Mamaindé "driveway" where translator David Eberhard met them in his four-wheel-drive. The track was another two hours of an often vertical, sometimes horizontal, always bone jarring backcountry boogie. Jon's job during the drive was to keep the two dozen eggs in his lap intact.

David's ancient Land Cruiser ground to a halt outside one of the larger huts. Bouncing up a mountain jungle track in a beat up four-wheel-drive is something like living inside a baby's rattle while the toddler holding it plays carpenter. It's not conducive to conversation. When the engine gratefully shuddered to a stop, the silence was powerful, clean, and welcome.

David broke it first. "Guys, the Mamaindé are a nomadic people. I can't guarantee that any of them will stay around long enough to get this job done."

"Well, we're here," Jon replied. "Let's trust God and get to work." Then he peaked into the egg carton. "Hey, only one cracked!"

The first order of business was setting up a recording studio. They got permission from the chief to use the schoolhouse, and Jon and the team rigged some old bedding on the walls and covered the windows for soundproofing. Equipment setup was next: a sheet on the wall for a screen, the video and audio recorders and microphones all found a place under the tin roof, for which they were grateful. But even though the tin roof kept things dry, the rain – a thousand tiny hammers banging away a few feet above the microphones – often halted their progress.

Jon had joined VMS in 1997 and began scripting videos. Most filmmakers try to synchronize a new sound track with the moving lips of the actors on screen. But it's practically impossible to get a word for word translation of a film into a *new* language. So the VMS technicians use scene sync instead of lip sync, the narrative following the scene instead of the actual speech of the actors. Jon prepared a script in English for the translator to put into the tribal tongue, in this case scripting Campus Crusade's well known *The JESUS Film*. (David had finished the translation into Mamaindé prior to their arrival.) Then the team goes to the field to make the audio track. The tribe's people provide the voices of the individual characters. The translator, David in this case, reads a line from the script that corresponds to the scene in the video, and the tribal "actor" repeats that line into the microphone. This gets everyone involved and raises interest and attendance for the debut of the film or audio.

Things went along fairly well until the team tried to record the part of Satan. Up until that point, the nationals had been interested and cooperative. The voices of Jesus, the narrator, Peter, James, John, and all the other characters successfully became patterns of particles on magnetic recording tape. But the "particles" wouldn't cooperate with the voice of Satan, or so it seemed. David, typecasting like a high school drama director, had chosen the tribal shaman for the voice of Satan. When the time came to record the pivotal moment of Jesus' temptation in the wilderness, the audio wouldn't work.

Jon couldn't believe it. He had worn headphones while recording and heard a clear voice like the others they had done that day. But on playback the audio was full of scratches and hisses and pops and buzzes.

He started trying to isolate the problem. "David, would you wiggle the connection on that mic?" he asked. "No, that's good. Let's swap the mics out and try it."

They recorded again. "If you are the Son of God, tell these stones to become bread." Again, the sound came through clear in the headphones.

"OK, let's play it back," Jon said as he hit rewind.

Hisssss, crackle, "If you are...hissss, pop!...stones to become... buzz!...bread."

They checked connections, swapped microphones and mic cables, wiggled wires, and checked speakers. Nothing worked. To complicate matters, they did not yet have the software necessary to record directly to computer. Instead they had to record onto 8mm tape, transfer the audio to computer, edit it, then re-record onto a master tape. Jon was recording the voices on the audio portion of the tape. Each section of tape had been quality checked before leaving the States. But at one point a whole section, video, music, and narration, simply vanished. The shaman only had three lines. The team ended up having three separate recordings on three separate tapes to get all three lines recorded. It took half a day.

But that half a day was only half the battle. The audio track had to be migrated over to video tape so that it could all play together on the VCR. That night Jon dubbed the three tapes over to the computer, splicing them together so that it sounded like one seamless audio recording. David was amazed at the quality. At that point Jon decided to go ahead and move all of the audio they had thus far over to the video master. He turned on the speakers to monitor the transfer. Everything went smoothly until, "If you are…buzz… hissss, pop!...the Son of God…crackle…stones to become…buzz!... bread."

But this time the missionaries realized a chilling difference. The noises were in a different place every time. No matter how clean the audio tape was, they were getting noises on the video. They did all of their checks again and finally realized that the problem wasn't with the tape. If the noises had originated on the audio tape, they would have shown up at the same place each time.

Jon and Dave looked at each other for a moment. "Are you thinking what I'm thinking?" They both knew spiritual warfare when they saw it. They just hadn't expected it to show up in a video tape. "We need to pray."

JAARS and Summer Institute of Linguistics (SIL) missionaries are highly educated people. SIL boasts more PhD linguists than any university on earth. JAARS is a collection of some of the most skilled technicians, pilots, and problem-solvers on the planet. They are people deeply schooled in scientific techniques. One of their distinctive characteristics is the application of the latest technological tools and scientific methods for the spread of the Bible and its message. PhDs don't look for demons behind every bush. Pilots and mechanics are not superstitious people. Those who are do not live long in the bush. They are hard-headed realists trained to think through tough problems, work down diagnostic trees, and develop real-world solutions. They don't lay hands on video tape recording machines and pray for healing…unless they need to.

This time, they needed to.

For the third time, Jon began recording to the master tape. This time, however, he did not change a thing, not a cable or a connection or a wire. As it neared the portion with the shaman's voice, the team put their hands on the equipment.

"Father, we recognize that this is more than a technological battle," they prayed. "We know that you want the Mamaindé to hear the message of Jesus, and we know the enemy of souls opposes that. We ask that you restrain the forces of darkness opposing us now and make it possible for us to get a clear recording. Amen."

Jon let the recorder run. "If you are the Son of God, tell these stones to become bread." He listened intently for errant noises but heard none. "If you are the Son of God, throw yourself down. For it is written: He will command his angels concerning you, and they will lift you up in their hands, so that you will not strike your foot against a stone." Not a buzz or a whistle came through. "All this I will give to you, if you will bow down and worship me." The audio was crystal clear. The team looked at one another and smiled. "Thank you, Lord."

Now their excitement was building. They knew they were in a spiritual battle; they knew God was in it and they were making progress. After reading their parts for about a week, many of the villagers needed to check their fields and do some hunting. All but "Jesus" and the narrator left. The VMS team thought that was all they would need, but they had forgotten someone.

Jon looked up from the script. "We need someone to read Zacchaeus's part."

David stuck his head out of the makeshift studio – about a stone's throw from the village – and saw a little boy. "Please go to the village and ask any man who is there to come help us." Due to the work that needed to be done, they didn't think anyone would be available.

The boy returned with CB. CB was the village thief. He was also the chief's brother which meant that he could not be challenged. Moreover, he liked to throw his weight around, lording his position over the rest of the tribe. Things were about to get interesting.

Quite often the Scriptures cannot be literally translated into a native language. Paraphrases and syntax substitutes have to be used. Jesus' encounter with the corrupt tax collector Zacchaeus, familiar as a Sunday School story to most Christians, was totally new to the Mamaindé and a complete shock to CB.

It was the tax collector's confession of faith and repentance that confounded the poor man. The NIV translates Luke 19:8 this way:

"But Zacchaeus stood up and said to the Lord, 'Look, Lord! Here and now I give half of my possessions to the poor, and if I have cheated anybody out of anything, I will pay back four times the amount.'"

But to convey it in Mamaindé, David had to render it a little more directly.

"I am a thief!" said David and nodded to CB for his cue.

"He says I am a thief!" cried CB, pointing to David.

"No, CB, not that!" Jon backed the tape up as David explained to CB the exact line.

"OK, let's try it again."

David: "I am a thief!"

CB, pointing to the video: "That man on the wall is a thief!"

Little beads of sweat appeared on CB's brow. His eyes widened; his breath came harder. The four little words became a jungle mountain that took two hours to scale.

Then it was time for Zacchaeus's next line.

David: "Anyone I have cheated I will repay four times!"

CB looked near cardiac arrest. His face lost all color. His eyes darted back and forth. A howl of mistranslations followed, until in a heap of terrorized exasperation he whimpered, "Anyone I have cheated I will repay four times."

CB was done with his part of the recording. But God was not done with CB.

Every people group has its own myth for the afterlife. Knowing that the traditional Western invitation to "receive Jesus" would've been meaningless to the Mamaindé, David crafted one that incorporated their story as a doorway to the gospel.

"All the people know that when a man dies his spirit goes down a wide path until he meets a giant anaconda. The snake eats his spirit, and he is no more. But some go down a narrow path, and this path leads to a place of peace and beauty and plenty. Only the ancestors know this path. That is why our people call upon them whenever one of us dies. This is what our fathers have taught us. But this Jesus story tells us something different. This story tells us that the man Jesus knows the way to the narrow path, and he has shown it to us. We must listen to him before we die, and choose, and not follow the ancestors. Only then will we escape the giant anaconda."

The Mamaindé are an active, nomadic people. It is not unusual for them to miss great occasions like weddings of relatives or other ceremonies if they are hunting or farming. However, all of them showed up to see the film when it aired. But when the hero of the film died, they started to leave, thinking the story was over. David had to assure them it wasn't. They returned to their seats and sat there transfixed, as the man they had just seen die, walked out of his own tomb.

When *The JESUS Film* in Mamaindé *was* over, when Jesus' victory over death was clear, they sat still as stones, saying nothing, overwhelmed with the power of the message. These people, who had had the gospel in their own language for over twenty years and never responded to it, were hearing it as if for the first time.

Jon Caton and the VMS team loaded up soon after that and went back to the States. But before leaving they asked David to review the tape one more time to see if he could spot errors or problems. As he watched on a monitor, the matriarch of the tribe, the old chief's sister, appeared behind him, silently watching over his shoulder the last thirty minutes of the film, never speaking, never moving.

At the end she spoke quietly, "These are true words. But my ancestors never heard these words." Jon's heart filled with gratitude when he heard her comments. He and David both felt certain now that their efforts had not been in vain.

One month later Jon and Kim Caton were in the United States celebrating Easter. David and Julie were also in a Western church, away from the village for a while.

Unknown to them, the Mamaindé had a church service that day too, the very first in their village. They didn't know it was Easter. They just knew that CB wanted to teach them more about Jesus.

Epilogue

Later in 1998, Jon Caton was at his mother's bedside as the cancer that had ravaged her body took her last breath. His brother and sister were there, and they were all being brave. Inside he was grieving hard, but he knew his mom was safe and at peace at last. But he heard again the Mamaindé matriarch's words and felt her grief for her ancestors.

And God spoke to Jon Caton then: "This is what I feel when someone dies who does not know me."

Jon is back on the field helping to speed the Word to those still waiting.

Paul Westlund and friend

3

Wounded Sparrow

*Are not two sparrows sold for a penny? Yet not one of them
will fall to the ground apart from the will of your Father.
Matthew 10:29*

The letter from his supporter came as a shock. Paul Westlund,
YAJASI pilot in Southeast Asia, opened it to read this question: "Paul, where were you and what were you doing at 3:30 A.M.
Eastern Time on April 11?"

Paul worked out the fourteen-hour time differential and flashed
back six weeks from the date on the letter to April 11. He knew
exactly where he was. He would never forget that day in the aging
twin engine Piper Aztec YAJASI used for over water missions.

A pilot's mind has time to wander on a long flight. Once the
course is set, the engines tuned, and the airplane trimmed at cruising
altitude, his workload lightens and he has time to think. That's what
Paul was doing on April 11, five thousand feet above the ocean and
two hours out of his destination – scanning his instruments, scanning the horizon, and thinking. At forty-two years of age and with
thousands of hours in small piston powered airplanes, he was re-evaluating his own horizons. He was thinking about the future.

He was thinking about leaving the mission field.

Thirty minutes into his reverie and with no warnings on any of
the gauges, the number two engine yanked his thoughts back into

the cockpit. Paul heard a loud noise, felt the engine shudder once, and die. He knew from the telltale clang that it would not restart. Instinctively he checked the GPS. He was one and a half hours from the relative safety of his destination.

Training took over as he killed the autopilot, chopped the fuel to the right engine and feathered the prop. He only had one option: descend to an altitude where the air was dense enough for the plane to fly on one engine.

But how low was that? And would engine number one take the strain? The plane was loaded to its maximum weight. If he went by the book, he would operate it at full power and cruise at 1000 feet of altitude. But it was an older engine. He couldn't take the chance that it would fly apart under the load. Paul put the engine at 90% and descended, number two hanging like a dead man on the airplane's right shoulder, creating drag where it had once produced thrust. Paul had to keep constant pressure on the left rudder pedal, fighting to keep the wounded bird straight and the sink rate manageable. As the altimeter unwound through three thousand feet, he reviewed his ocean survival skills. Worst of all, Paul's life wasn't the only one at stake.

The small native man in the right seat reached over and lightly touched Paul's arm. "Are we going to make it?"

"Yes."

But the man didn't let go, instead, he put his fingers on Paul's wrist. Checking someone's pulse is the jungle's lie detector. Paul's heart was already drumming in his ears – it couldn't beat any faster – and he knew his pulse would reflect that. As they passed through two thousand feet, Paul started thinking about surviving, period. The man asked him again, "Are we going to make it?" and felt Paul's pulse. One thousand feet and still sinking.

"Yes." Once more he took Paul's pulse. But the descent had slowed along with his heart.

At six hundred feet above the waves, low enough to read the name on a cargo ship he passed, the old bird leveled out. They might make it.

For the moment, Paul left those memories behind. Someone closed a door behind him, breaking into his reverie and jerking him back into the present. He looked back down at the letter in his hand, the one from his supporter, asking where he'd been. At 3:30 A.M. Eastern Standard Time on that April 11, 1996, Paul was holding a half-dead airplane six hundred feet above the South Pacific and thinking, *If I get out of this alive, I'm going back to the States.*

To the young, serving God in the jungles or giving one's self to full-time ministry of any kind looks like a grand adventure. But twenty years into it, the perspective changes, even for missionaries. Paul realized that retirement was coming, and his nest egg was nowhere near the size of his friends' in the secular world. The strains of life in the fishbowl, always dependent on the goodwill of churches and donors, always forced to wear cheaper clothes, drive cheaper cars, and take cheaper, shorter vacations, start to grind. Life seemed tougher than it had to be, and God seemed farther away than he was when Paul was young.

In his more honest moments, Paul knew that was what drove his fantasies of returning to the States. It wasn't the money or the equipment or the strain of flying over trackless jungles and open water. What Paul really needed wasn't money or vacations or an expensive car. It was a touch from God.

It came in the letter he was holding in his hands. It read:

"Paul, as you know, I live alone. At 3:30 in the morning of April 11, someone slapped me on the back of the neck. I woke up immediately, wondering 'what in the world?' but saw nothing. Just then I became intensely aware that you were in trouble over water, and I needed to pray for you. I prayed. I could not stop praying for an hour and a half. Paul, are you OK? Or am I just an old woman having strange dreams?"

Paul leaned back in his chair, gentle warmth filling his soul like sunshine on a spring morning. His eyes glazed and began to fill with tears. He remembered that at that exact time he'd seen a small island beach from six hundred feet above the sea. The tide was out. He could put his wounded bird down there. He knew he could make the landing. But the airplane would be a total loss as soon as the tide came in. He pushed on another half hour or so. The crash trucks had been ready. They'd heard his distress call, but he didn't need them. Carefully, gingerly, with fifteen years and thousands of hours of skill, Paul put his ailing bird back on solid ground. Six weeks later, looking at the letter, he knew how closely God had been watching his wounded sparrow.

Epilogue

Sometime later, Paul heard from a friend named Scott. Scott flew corporate jets for a big tool company in the States: Gulfstream G-IVs, King Airs, all turbine stuff with pressurized cabins, the latest avionics. These planes are the Lexuses of the air. "I don't want to steal you from the mission field, but if you are ready, I have been holding a jet job for you," he wrote.

A pilot usually has to have one thousand hours of airtime in turbine powered planes before a big outfit will consider him for such service. All of Paul's time was in small piston powered planes. Scott was opening a huge door for him with great pay and benefits, one that didn't depend on thirty-year-old airplanes, short mountain airstrips and dances with destiny above the waves.

Paul knew what to do and answered, "Thanks, but no thanks. I cannot trade the prayers of God's people for the coolness of flying a jet and the money it would bring."

Something to replace the venerable Helio Courier

4

Flying on Empty

❦

Providence is the practical outworking of the will of God in the lives of men that appears from our perspective as tragedy, chance, or circumstance.—Matthew Henry

"We need new airplanes. We need *seven million dollars'* worth of new airplanes. And we don't have a dime."

No one remembers who the first person was to say out loud what everyone was thinking. But it summed up where the Southeast Asia JAARS leadership found themselves at the turn of the twenty-first century. A bold new vision was flowing through all of the Wycliffe organizations: Vision 2025. The mission had committed itself to trust God for the startup of a Bible translation work in *all* of the three thousand as-yet-unreached people groups of the world by 2025. The lead pilots and mechanics in Southeast Asia, where JAARS operates in partnership with a national organization called YAJASI, knew they could not service the needs of the translation teams that would pursue that vision without new airplanes.

The venerable Helio Courier, a robust single engine, six passenger flying "Jeep of an airplane" had been the backbone of the YAJASI fleet for over thirty years. Its qualities as a bush plane were unmatched by similar birds. But two things were rapidly pushing the Helio out of the hangar and into the boneyard: *age* and *avgas*.

JAARS knows more about maintaining the Helio than any other bush flyers in the world and operates the largest fleet of the unique plane with its automatically deploying Handley-Page slotted leading edge. Its maintenance standards and procedures are unsurpassed. But the fleet average of seventeen thousand hours is a long time for any airframe. The dangers of pushing past that are well documented. Since the Helio is no longer in production, JAARS had to make a lot of its own repair parts. This kept the planes grounded more and more often as mission needs took a back seat to safety. Still, they probably would have kept at it but for the global shortage of avgas.

Avgas is much like auto gas, only with higher octane ratings, and everyone knows what has happened to the price of gas. But where JAARS flies, price, even at twelve dollars a gallon, isn't the only factor. Often the fuel is simply unavailable. Jet fuel on the other hand, the fuel used by airliners, is more like diesel fuel and more readily available in the developing world.

One of those most familiar with the Helio Courier and the need for a replacement is JAARS / YAJASI pilot Nate Gordon. At forty years of age, Nate is of average height, very slim, and mostly bald. He sports a mustache and goatee, trimmed short as is what is left of his still dark hair. He is affable if not exactly easygoing, a focused man, but one who lacks the hard edges often found in such intense people. He smiles easily and, without affectation or whispery mysticism, is completely intoxicated with God, a serious minded follower of Christ.

Nate was born in India and grew up there, in Nepal, and the Philippines. He has been flying for YAJASI in Southeast Asia since 1995. As YAJASI Aviation Manager, responsible for day-to-day operations, he knows the situation intimately.

With over one hundred and twenty plus languages left to translate in the region and with every language group needing air service, Nate and YAJASI knew they had to replace the seven Helio Couriers currently in service or the task could never be done. So they began to dream, "What airplane would do the job?"

In the field, JAARS operates as a small air service for their host countries. They haul supplies, passengers, business and government agency representatives and such for a fee. The money they earn doing that offsets costs for the translators they carry. It can't cover all of it, but it helps. This meant that any replacement for the Helio had to be certifiable by any country's aviation authority as a commercial aircraft, thus ruling out experimental types.

Additionally, the plane had to be turbine powered, already in production and very likely to remain in production. Factory support was a must. JAARS had to be able to get parts. And of course the planes had to be very sturdy, able to operate in the mountain jungle environment with short runways. YAJASI needed a flying truck as rugged as the Helio Courier that could run on jet fuel. Since the Quest Kodiak was not yet in production, the turbine powered Pilatus PC-6 became the obvious choice, the only choice really, but it is a one and a half million dollar airplane.

YAJASI was replacing four Helios and two twin engine Piper Aztecs used for flights over water and service to longer airstrips. They needed five of the larger PC-6s, seven and a half million dollars' worth. They were asking for more money for a single field project than any other venture in the history of the organization.

When Nate and other YAJASI members started talking about it with mission leadership, they got incredulous looks. "Do you really know what you're asking? Here's how big the fund raising pie is. You're asking for more of it than can be spared."

But Nate and his YAJASI friends had a more faith stretching idea. "Well, we don't think we're asking for more of the same pie. We think we're asking for God to give us a *new* pie."

It took some doing, but by late 2002 they had permission to start the process of finding that new pie. A special fund was set up, brochures went out, a website went up, and after eighteen months they had "zip." Nate had been flying in Southeast Asia for over a decade and had lots of stories of God's provision and intervention. But this time, he could not foresee how the need would be met.

Still, he and the YAJASI team began praying regularly for the new airplanes. They had done their due diligence, sending out fund-raising letters, setting up a special account for the project, advertising the need wherever they could. But they had nothing. Things kept getting worse and worse on the field. Finally, YAJASI ran out of avgas; they literally could not buy enough to sustain normal flight operations. Without gasoline they couldn't fly the paying passengers. YAJASI went into the red for the first time and had to ground the fleet, sometimes for as much as a month at a time, waiting for the next shipment of avgas. Vision 2025 dropped off the horizon. YAJASI struggled along in this situation for the next three years, sometimes able to fly, sometimes grounded except for essential mission flights or emergencies.

Enter a Dutchman known only as "Paul" to protect his anonymity. None of the YAJASI people knew Paul. He was in Papua on a totally different errand, and he didn't know about YAJASI, who they were, what they were doing, nothing. Paul is a Christian businessman with multiple successful ventures in software and aviation. His charitable mission is to help impoverished Christians the world over develop economically in God-honoring ways. He was in Southeast Asia looking for ways to accomplish this. Paul had a field leader in Papua named Nico, and Nico had heard about YAJASI. When his employer arrived in Papua, Nico told him, "You need to meet these YAJASI guys."

Nico set up a meeting between his boss and a few of the YAJASI people. Nate slapped together a PowerPoint, a few photos, videos and talking points, and went to make a thirty-minute presentation.

"Here's what we are at YAJASI," he said as he scrolled through the photos. "Here's what we do. Here's what we think the future looks like. Here's what we think we need." That was pretty much it for Nate's presentation.

Thirty minutes turned into two hours as Paul kept plying them with questions like "What people groups do you serve? How many planes do you operate? What kind of loads will you carry? How often are you able to transport goods to market?"

Nate had no prior history with this man – they had never met before that day. But at the end of Nate's presentation, Paul turned to his chief pilot, who had accompanied him to the presentation, and in Dutch asked, "Does what this guy says make sense?"

"Yes, he's exactly right," the pilot nodded.

Nate could sense the presence of God in the encounter. Paul revealed that his son was also a missionary in the region. He validated what Nate was saying, and though he didn't know YAJASI he could speak to the cultural environment. So, unknown to Nate, there was built-in credibility and believability. Nate believed all this was the handiwork of God, but had no idea what would come of it.

As the meeting came to a close, Paul shook Nate's hand and said, "Thanks, it was nice to meet you," and left.

Three months went by. Nate had almost forgotten the incident when he happened to overnight in another part of the island. He went with another Wycliffe colleague to a social gathering and there, ran into Paul's assistant Nico who said, "I need to talk to you."

They found a quiet corner, and Nico said, "We want to buy an airplane for you. We want to buy you a PC-6."

Joy filled Nate's soul at that moment, joy suffused with the light of conviction that God was indeed working to supply the needs of his servants, to equip them to keep taking his Word to the forgotten peoples of Papua, to "give them a new pie." But he was still surprised. "Really?" He smiled and almost laughed as he said it.

That first PC-6 arrived in Southeast Asia in 2005. Working with YAJASI'S chief pilot, Tom Beekman, Nate's task that year was to initiate "Paul's" donated Pilatus PC-6 "Porter" into service.

A couple of Helio Courier flights in the mountains and anyone would understand why the missionary pilots love the older bird. It's rugged, predictable, and gets off the ground and up to altitude faster than anything in its class. But one trip in the PC-6 showed Nate the wisdom of their choice for a replacement. Takeoff is strong in the Helio. It's like riding an express elevator in the Porter. The Porter hauls twice as much cargo, climbs faster, can land shorter and fly higher than the Helio, and do it all more reliably on jet fuel.

That was PC-6 number one. With it, YAJASI moved back into the black. They would be even more astonished to find out that God already had a second one in the works.

In July of 2005, as a second PC-6 was being removed from its shipping crate, assembled and equipped by YAJASI personnel, a man in New York moved his mouse across the pad and brought his computer to life. He clicked on one of his favorite sites, pc-6.com, and read the news of the latest Porters to come off the line. Nate didn't know it at the time, and neither did any of his YAJASI colleagues, but Marshall Carter was watching the progress of both planes. As one of the few Americans with a PC-6, he had a personal interest in the aircraft and how they were used. When a new one came off the line in Switzerland, the site posted its serial number and destination, functioning as a kind of community database for PC-6 operators worldwide.

Marshall saw the first YAJASI PC-6 on the site. Then he saw a second one going to the same place. *Who are these guys?* he thought. He would have been even more intrigued if he'd known where that second plane came from.

The story of the second PC-6 is perhaps more remarkable than the first. In April of 2004, a phone rang at Wycliffe headquarters in Florida. A youth pastor was on the line with an unusual question.

"Do you have any current projects that could use a million and a half dollars? You see, my wife and I have recently come into a large inheritance. We know that this much money this early in life could destroy us. We don't want that to happen. So we're giving it to the Lord's work."

And just like that a second PC-6 was on its way to YAJASI.

"Materialism isn't how much you have," Nate once said. "It's what's number one in your heart. You can be dirt poor and be ready to throw your integrity out the door to get a buck and you are a materialist. The people I serve, though very poor, are just as afflicted with it as Americans. That's what makes these airplanes, the way we got them, as powerful as anything we do with them."

Nate makes it a point to tell his tribal friends who depend on YAJASI services, "Do you see this new airplane?" He'll be sitting on a jungle runway with twenty or thirty tribal guys around. "It was given to you by a poor man. The other one was given to you by a rich man. God uses anybody who will honor Him. Imagine yourself being given one and half million dollars. You are set for life. You'll never have to work again. And you think of people on the other side of the world, whom you will never meet, who have nothing. And you give that money away so that they can know God's love. That's why this airplane is parked on your runway today and flying your sick family member out or your medicine in or your pigs to market."

Half a world away, Marshall Carter started to wonder about this outfit called JAARS. He clicked on the jaars.org link listed by pc-6.com and started to study them and thought, *This is pretty good stuff these guys are doing.* He picked up the phone and called JAARS headquarters in Waxhaw, North Carolina.

"You guys don't know me but I own a PC-6 in the States," he said. "I've had some modifications done to it and thought you might like to see them. I've worked through the certification process with the FAA. I'd love to be a resource for your guys over there in Southeast Asia as they bring them online. Here's my contact info."

It was just one aviation guy connecting with some others, not uncommon. None of the JAARS folks knew anything else about him. So the technicians at the JAARS Center in the States sent YAJASI an email and said, "Hey, we don't know if this will help, but you might want to talk to this man."

Later on Nate and the YAJASI technicians did come across an issue on the PC-6. They needed a minimum equipment list for the airplane and thought, "Let's give this guy a call. Maybe he's already got one of these we could use as a template."

Carter, Marsh to his friends, was more than helpful. He boxed a beautifully prepared and bound version of the equipment list and shipped it to YAJASI. It was exactly what they needed.

Not long after that, Nate was on furlough and planned to visit family near where Carter lived. He'd been so helpful that Nate wrote him and said, "I'm going to be in the area. Can we get together?"

"Sure," Marsh replied, "we'll go flying." So the two men met, but the weather was zero, zero (zero visibility, zero ceiling) so they couldn't go flying. But they did have lunch in Carter's hangar talking airplanes while Nate, who'd never seen an amphibious version, inspected Marsh's PC-6. They were just two aviation guys talking shop.

Nate had his laptop with videos of what YAJASI does on the field and said, "Hey, since we can't go fly would you like to take a look at what we do?"

"Sure," Carter said. Nate fired up his computer and let it run, Marsh asking questions and Nate filling him in as the videos played. They got to the end, it was time for Nate to go, and Carter looked at him for a moment and said, "Man, I wish I had spent my life doing what you're doing instead of what I'm doing."

Nate said, "What do you do?"

"Oh, I've been a bank president a couple of times. I'm retired now, but it isn't working out for me."

"Oh, how come?"

"Well, they're asking me to be the president of the New York Stock Exchange Euronext."

Nate was more than a little surprised. Marsh was very down to earth and unpretentious, not what one would expect from recent news portrayals of Wall Street executives. But then, it wouldn't be the first time the media got it wrong.

As Nate prepared to leave, Marsh shook his hand, looked at the missionary pilot, and said, "You know, I'm not going to have my medical [airman's medical certificate] much longer. I'm sixty-three and something will come up. One of these days I'll lose it and have to stop flying. When I do, I want you guys to have this airplane."

Nate didn't know what to say. He thought back to all the connections in the story, the fact that JAARS did nothing to contact Carter, the fact that another man they didn't know, a Swiss gentleman, put JAARS on his website, that Marsh happened to see the link to JAARS, clicked it, and now they were being offered a third airplane. Nate felt the same way he did at the end of his conversation with Paul the Dutchman. A powerful conviction was growing in his heart: *God wants us to know that it's him, not our fund-raising, that's doing this.* Stunned but grateful, Nate said, "Thanks, Marsh, that means a lot to us," and left.

YAJASI now had two new airplanes in operation. They didn't know if they should try for the third, but JAARS leadership agreed to go ahead with the audacious plan to trust God for three more. So they wrote Marsh Carter and said, "We just want to kind of see where you stand with that offer because it will impact where we go from here." In other words, YAJASI needed to know if they were raising money for three more airplanes (figuring they needed five to do the job) or just two with the promise of a third coming somewhere down the line in a donation. It was a polite way to say to Mr.

Carter, "Now that you've had time to think about this, are you sure you still want to do it?"

Marsh wrote back and said, "Yes. I'm still committed."

YAJASI wrote back and said, "OK, great, now we know what we have to plan for." So they started to work on funding the fourth airplane.

Two weeks later Marsh Carter wrote back, "I've changed my mind. I want you to have it now."

The third Pilatus PC-6 given to YAJASI, Marsh Carter's plane, is the first pre-owned bird and the first one to be shipped from the United States mainland, not from the factory in Switzerland. It had to be inspected and modified – floats removed, complete mechanical inspection, and new avionics installed (notably a Traffic Collision Avoidance System), before it could be shipped to Southeast Asia. Late in 2008 Nate returned to the JAARS Center in Waxhaw, North Carolina, to make the maiden "return to service" flight that initiated Marsh's plane as a member of the JAARS fleet.

As he flew, Nate reflected on all that God had done to bring four and a half million dollars' worth of airplanes to an organization that had no money to buy them.

We did nothing to contact these donors. They found us. An acquaintance mentions us to his boss. A couple calls Wycliffe out of the blue. A Swiss PC-6 enthusiast puts us on his website and Marsh Carter just happens to click the link. He made a turn and headed back to base. *God wants us to know "I did this,"* Nate mused. *It had nothing to do with slick fund-raising efforts. God is saying to us, "You guys are right on. If you will be true to your calling, if you will not waver from reaching the least of these, the forgotten people on the planet, with my love and making sure that they have my Word in their language, I'll make sure you have what you need to make that happen."*

And because of all that, like a pair of far off wings glinting in the sun, Vision 2025 is back up on the horizon in Southeast Asia.

As he turned on final, Nate's face broke into a grin. *I can't wait to see how God provides for the last two.*

Epilogue

Nate Gordon continues to serve YAJASI as Aviation Manager. His friend and fellow aviator Paul Westlund now flies the PC-6 every day in service to the people of Southeast Asia.

Nate Gordon and PC-6 at village

PC-6 power launch

After landing it's often an uphill climb to the loading zone.

High altitude bus stop

End of an era engine. Tim Harold works the
final refit on a Helio Courier.

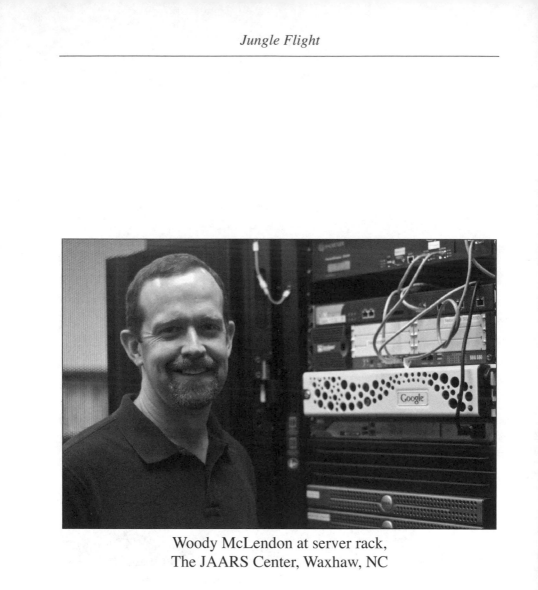

Woody McLendon at server rack,
The JAARS Center, Waxhaw, NC

Cessna 206 departing village in Papua New Guinea

5

I Can't Stay Here

So whether you eat or drink or whatever you do, do it all for the glory of God. 1 Corinthians 10:31

Chuck Daly gazed out over the concrete ramp at the JAARS base in Ukarumpa, Papua New Guinea, and, with pen poised above his log book, flashed back twenty years. He was a kid, growing up with missionary parents, SIL translators in southern Mexico. Small planes like the one he had just landed had been the only way, short of an eight-hour hike through the mountains, to travel in that part of Mexico. He remembered the sound of the Mission Aviation Fellowship plane droning overhead on approach, the hollow rattle of the airplane's aluminum panels as they flexed and strained, landing on the rough dirt strip on top of a mountain in Oaxaca. Most of all he remembered roaring down that runway and zooming into the sky with the friendly pilot, who brought them everything they needed and helped them serve the Mixtec village they called home. He'd known right then, just a boy of eight or nine years old, that he wanted to be a pilot. Now he was living his childhood dream.

But there was something wrong, a disturbance in this vision, a distortion in his heart whose roots he could not discern. Daly, a ruggedly handsome man with close cropped hair, and the quiet, steady manner of an experienced aviator, thrived on challenge. He was not accustomed to problems he couldn't solve and he didn't like it.

A pilot calling, "Clear prop!" and cranking up nearby brought Chuck back to the moment. "Lord, I love my job!" he prayed. "But I'm pushing everybody, especially Chris, my wife, really hard. Please help her to hold up." He finished his log entry and headed for his bike.

Chuck and Chris Daly had arrived in Papua New Guinea (PNG) in 1992, so Chuck could serve as a line pilot flying the Cessna 206. Chris had been pregnant with their second child, and life on the equatorial island in the developing world was rapidly wearing her down.

At first, they were living in a place called Nobnob, a little village near Madang on the eastern coast of the island nation, a few degrees of latitude north of Australia. SIL PNG had a training center there, a kind of "this is how we do it in the developing world" initiation program designed to equip new arrivals for the challenging life they had chosen. The regimen of swimming and hiking was hard on a woman in the early stages of pregnancy. But physical exercise wasn't the only problem. Boiler-like equatorial heat and humidity, life in rudimentary dorms complete with paper thin walls, communal bathrooms, flies, malarial mosquitoes, and cockroaches burned through Chris's emotional reserves. The living conditions, her three-year-old Matthew's needs, her neighbor's children, and the minimal privacy offered by the thin walls and open windows had robbed her of precious sleep. She was worried about the child she was carrying, and struggling spiritually. *Why, if God led us to this place, is it proving so difficult?* she thought. Her spirit continued to persevere but her body could not.

Late one evening she touched Chuck's arm. "I think I'm having Braxton-Hicks," she said, mentioning the early contractions some women have before they are ready to give birth. A checkup confirmed it. The stress was pushing Chris into early labor, putting her and their new baby at risk. They would have to pull out of the training program early.

That was the day that Chuck's admiration for his father took a step up. He had always loved and respected his dad. But he began to realize in Papua New Guinea that it is one thing to grow up on the mission field and love it. It's something else to be responsible for the lives of other people who are with you, to be the husband and father. For himself, he was beginning to feel like a plane loaded to the limits of its center of gravity, finding it harder to balance the burden.

The abbreviated training didn't mean they had to go home. It just meant that they went to Ukarumpa – the inland SIL base – with a black cloud over their heads, a sense of failure and uncertainty. The strain caused confusion about what God was doing and it came out in their prayers. "God, you clearly led us to PNG. Why are we having such a hard time? We expected life to be good out here and it's not good. It's hard and we don't understand."

Chuck and Chris prayed like that for two years with no clear answer. Little Alaina was born healthy just two weeks shy of full term. Chris began to adjust to the strain of living in the developing world. But the young couple, now in their late twenties, still struggled with something nameless, some disquiet in their lives, until one day in 1994 a light switched on in Chuck's heart.

He stood in the hangar, elbow resting on the edge of a toolbox, talking to one of the mechanics. A memory, not quite a déjà vu, came flooding into his consciousness. In his memory, he was in a similar pose in the JAARS hangar at Waxhaw, North Carolina, talking to fellow pilot Carlos Rangel. That was in 1991 and they were in JAARS flight training together, absorbing the lessons that would prepare them for jungle aviation.

Chuck remembered saying something like "Isn't this great?" Carlos's response rocked him way back then. It hit him now like a thirty-knot crosswind on final.

"Chuck, I don't care if I ever fly again, as long as whatever I'm doing glorifies God and helps the kingdom grow."

The answer to two years of prayer came crashing into Chuck's consciousness at that moment in the Ukarumpa hangar. "I've been doing all of this," he said to himself. "I've pushed my family to the breaking point for *my* dream, *my* glory, not God's. I love *jungle flying* for God more than I love God."

Stunned by the revelation and needing some space to think, Chuck walked out on the ramp and scanned his surroundings. Ukarumpa is in a high mountain valley, surrounded by rolling hills of vivid green, comfortably breezy compared to the sweltering coastal plains, a beautiful place. He gazed long at the late afternoon sky, brilliant blue against the lush and towering cloud-scraped mountains in the distance, and felt the warmth of the day beginning to fade. He took in the runway stretching far, far down the valley. The windsock on the hangar roof barely moved in the gentle breeze that caressed his face. He drank in the airplanes parked nearby, even the smells of avgas and oils lingering in the air, things he had known and loved all of his adult life. *I have to give this up,* he thought. *I have to let this go.*

As powerful as the conviction was, it still took Chuck two months to tell his wife. They had sacrificed so much to be in PNG; he did not know how she would take it, and he wanted to be sure in his own soul before he said anything. But the day came when he walked into their small home in Ukarumpa, took her by the hand, and said, "I need to tell you something." Chris could see that he was struggling deeply. She looked him in the eyes and waited.

"We can't stay here," he said. "You know how we have been praying for the last couple of years, that God would cause our hearts to be united again, wanting the same thing?" She nodded. He went on in his quiet but steady way. "I know what's wrong now. Jungle flying for God has been my idol. My dream of mission flying, my personal glory in it, has been more important than his glory. God showed me that so clearly that I can't stay here and keep doing this."

"Are you sure, Honey?" She looked lovingly at him. "Because you know, I'm okay now. It was hard at first, but God has given peace in being here and I'm fine with staying."

Chuck smiled. He was so grateful for this woman. "Chris, God wants me to give it to him completely. Anything less, if I stayed here and kept on flying…anything less would be a failure to acknowledge the reality of what he has shown me. I'm sure."

"Then let's go."

Telling Chris was hard, but explaining his decision to his SIL colleagues would be harder still. It was not that they did not appreciate the spiritual dimension of Chuck's decision. JAARS and SIL people are deeply spiritual, serious about following Christ. But they had seen what the stresses of living in the developing world could do to a young family and knew that an early exit from the field was often a great temptation.

"Chuck, I understand the desire to please God and I honor it." His supervisor, Don Archibald, looked across his desk as the young pilot finished his explanation for leaving. "But let me ask you to reconsider. You've got so much invested here in money and time and frankly so do JAARS and SIL. All of those hours of training in Waxhaw, the field training here, and check-rides on all of the airstrips, it has to start all over with a new pilot if you leave now. Are you sure you've understood God's will clearly?"

Chuck nodded. No one loved flying in Papua New Guinea more than he did. He loved the challenge of aviation in the toughest conditions on earth, he loved the high performance airplanes they used, and he especially loved delivering God's Word to people who had never had it in their own language before. He did *not* want to leave. But the conviction was too deep, the issue too clear, and he knew that it would be wrong to stay. "I have to go."

As much as he loved it, as much as his life had been shaped by it, Chuck did not know if he would ever fly again or even if he would stay with Wycliffe.

The Dalys were soon deep into preparations to return with their children to the States, too consumed with the multitude of tasks required to think much about what they would do next. Not until they reached Tucson, Arizona, settled into temporary quarters, and began making visits to their supporting churches did Chuck's prayer life take on a new dimension.

"Lord, I want to do whatever you want me to do, whether it involves Wycliffe or flying or neither of those things. I want what you want for me, not my will this time, Lord, but yours be done." It was the first time he had ever prayed that prayer. He was beginning to learn that it would not be the last.

Glen Mast, then JAARS aviation training manager, had recently written to Daly. "Chuck, I'd like to ask you to consider something. We need another instructor in the training program. You've been through the program and had two years on the field as well as a lot of instructing experience prior to joining JAARS. I think you'd be a good fit."

Chuck agreed to pray about it. As much as he had loved jungle flying, he had grown tired of instructing. His mind returned to long days in the Arizona flight school where he had earned his wings and then, to build time, stayed on as an instructor. *Sweltering in unairconditioned cockpits at low altitude, constantly on watch for collisions in the heavily trafficked air near Tucson is not my idea of a good time,* he thought. He remembered the boredom interspersed with moments of terror when he had to decide whether to let the student make mistakes or intervene. Too much control and the student couldn't learn. Too much leeway and the student could tear up a valuable airplane in one of the repeated takeoffs and landings they had to make. Training pilots for the mission field would include all of that and more. It was not, in Chuck's opinion, a shining option.

But it *was* a critical contribution to Wycliffe's mission, that much had become clear in the Dalys' prayers. God had taken Chuck off the line as a jungle pilot and brought them back to the States, but he had not released them from ministry through Wycliffe. They still felt the burden to be part of the mission to make the Scriptures available in the mother tongue of people who had never had a Bible. And JAARS still needed well trained pilots.

Once again Chuck prayed, "Lord, this is about you, not what I want, not my will..."

Daly was soon on the flight line at the JAARS Center, training others in the art of jungle aviation. He stayed in Waxhaw until 1998 when word came that JAARS needed a pilot to stand in for another aviator on furlough from the program in Cameroon, Africa. Chuck took his whole family to Africa for four months and once again lived his dream. Only this time they could feel the Lord's blessing in the service, almost as if he were saying, "Because you have honored me with obedience, I'm giving this back to you for a while." It was a sweet feeling.

From Cameroon Chuck returned to the training program in Waxhaw. Then in 2001 it became clear that JAARS needed to shut down one of its oldest aviation programs in Peru and turn the work over to South America Mission (SAMAIR). The task would require a lot of flying and last about two years. Again, the Dalys packed up and moved out of the country. And again, the flying felt like a gift from God.

Little did the Dalys know that not only was Peru a gift to them, Chuck's flying would be a special gift to a little Indian boy named Onésimo.

Chuck Daly in cockpit

6

Onésimo

Your neighbor is anyone whose need you see, whose need you are in a position to meet.—Haddon Robinson

July 2003, Yurua River, Eastern Peruvian Jungle:

Onésimo's momma, Shuro, frowned as she called after her little four-year-old boy, "Do not go, Son, Puma'll get you," she said, stretching the *u* in Puma for effect. The little boy frowned, his short, dark hair framing brown, almond eyes. But frowns did not stay long on Onésimo's face. Two and a half feet tall, with a deep copper complexion, the little Chitonahua Indian boy was easy to love but hard to guide once his mind was made up. He smiled and kept walking, barefoot like his father. Down the jungle trail he went, toward the small lake formed by a river they would canoe across before continuing to the hunting ground. Onésimo's sister Yapa, seeing he wasn't listening, nodded to her mother as the older woman tried to soothe a fussing infant at her breast. Calling out again over her shoulder, Shuro cried, "Do not go with your father on the hunt! A spirit will bite you!"

All mothers with four-year-old sons say such things. And all four-year-old sons ignore their mothers.

The puma didn't know this, but his strategy depended upon it. In the Amazon he is not a figment of Mom's imagination. He is

180 pounds of muscle, claw, fang, and stealth, *Panthera onca*, the jaguar. He is camouflaged like the leopard, only with bigger, thicker rosettes against a darker yellow coat – so dark in some that he easily fades to black – perfect cover in the light-squelching, triple-canopied jungle of Peru. His jaw is the most powerful of the big cats. Only five feet in length, he can drag a six hundred pound bull one hundred feet into the jungle and break its bones with one bite. Thus to the Chitonahua he is the puma, meaning "beast that kills with one pounce."

The puma crouched low in thick undergrowth that day, stalking little Onésimo, his tail swinging ever so slightly from side to side, as the boy trailed his father, Whunani, on the hunt. Just before a bend in the trail that would take him out of sight, Onésimo turned, waved to his mother and smiled. Then the jungle hid him from view.

Ten-year-old Yapa, Onésimo's sister, had considered the situation and made her decision. She would return to the village. Her brother was going with father on the hunt. Her mother, Shuro, was tending to a sick toddler. There was nothing for her to do but bundle her belongings in her mosquito net and head for school. She picked up the net and walked toward the trail that Onésimo had just followed.

The puma's tail stopped swinging, every muscle on his compact body coiled for attack, all of his senses focused on the copper colored boy with the smile on his face. Then silent and quick as death the cat lunged, covering the twenty feet to his prey in a second. Onésimo's head was in the puma's mouth before the boy even saw him.

"Aiiee! Come quick, come quick!" Yapa screamed. "A puma is eating little brother!"

Yapa's route had taken her down the same trail and toward the lake that Onésimo was skirting. The cat had pounced as her brother came into view. Her screams set Shuro's feet on fire. She too screamed and cried as she ran, still holding her toddler, reaching a hill overlooking the lake just as the jaguar adjusted its hold. Sharp fangs pierced the boy's face and skull, and the cat began dragging him toward the jungle. Shuro's screams reached Whunani, now one hundred yards down the trail. He turned around and began running toward the sound.

"Stop! Let him go! Let him go!" Shuro and Yapa screamed at the puma, fighting back fear with courage. Then suddenly, before Whunani had made it back to the scene, the cat dropped Onésimo, grabbed the small bag the boy had been carrying, and ran off into the jungle.

Whunani arrived to find his son a bloody mess. He had two holes in the back of his skull. His left cheek was ripped open. There was a round puncture hole in the bridge of his nose, at least 1/2 inch in depth, with pieces of broken bone visible. The puncture angled in toward his left eye which protruded out of its socket. His right cheek, eyelid, and lip were lacerated and all of the wounds were bleeding profusely. Without a word he picked up his son and ran back down the trail toward his canoe. Shuro handed her baby off to Yapa and followed him.

"We will take him to the medical aid station at the military base downriver," said Whunani as Shuro cradled her boy in her lap. Onésimo screamed in pain. Whunani shoved the little craft into the water, jumped in, and began to paddle. The trip, he knew, would take all night.

The Following Morning at a Jungle Airbase, Pucallpa, Peru:

JAARS Pilot Chuck Daly was in the middle of a day that began at 5:00 A.M. He was there to close a JAARS hangar that was no longer needed, a thousand details swimming in his head and no knowledge of the drama unfolding one hundred miles away on a lonely river. Chuck's mind was wandering, wishing he could be some place other than this boring meeting. He came to Peru to fly. He was about to get his wish.

"Mr. Chuck," a tiny woman who served as secretary for the base interrupted with a worried look on her face. "There is a call for you. Someone needs an emergency flight." Chuck bolted for the phone and then for a car. Forty-five minutes later he was on the runway, in a Cessna 206 fueled and preflighted by JAARS' partner SAMAIR (South America Mission) personnel ahead of his arrival. An hour later he was on the ground at the aid station in Puerto Breu. He had already been praying, but now his prayers took on new intensity.

Chuck heard the boy's screams before he saw him, and when he came into view, Chuck realized he had never seen anything like the gory victim in his mother's arms. They had immediately put Onésimo in a canoe and headed downriver to the aid station. But that was an eight-hour trip. The attack had happened almost twenty-four hours before Chuck arrived, and the boy had lost a lot of blood. He was screaming in pain. Chuck didn't think he would make it.

It was a nightmarish scene, Onésimo's screams filling the air around the airport and following them into the Cessna. Shuro held her son in her arms and the nurse walked beside her out to the plane, holding an IV bag attached to the needle in his arm. The jaguar's lower fangs had punctured the back of Onésimo's skull in two places. The upper fangs had gone through his septum and the bone under his left eye socket. His left eye was punctured, creating a hematoma in the eye socket that was pushing the eyeball out onto his cheek. Blood seeped through the bandage on the back of his head. But there was no bandage on his face. The wounds were gruesome and the child could not stop screaming.

Chuck had rigged the stretcher used for medical evacuations. Normally, no medical patient/passenger is allowed to ride cradled by another person. Heavy turbulence or a crash landing might injure them or throw them around the cockpit. But one look at his mother told the pilot she would not let him go. He strapped her into her seat, and put a child's lap belt around Onésimo while she held him. Then he hung the IV bag from a seatbelt clip on the ceiling.

As he turned to climb into his own seat, Chuck saw something that shocked him. Shuro's husband was climbing into the airplane. Chuck didn't know what he was doing, and thought he was trying to go with them. But there was no room, and he was about to chase the man out when he realized what was happening. Onésimo's father was hugging them and saying goodbye to his wife and little boy, an intense display of affection Chuck had never seen before or since in a Peruvian Indian.

Daly was praying hard, as he configured the plane for takeoff, that the boy would not die. He taxied down to the end of the runway, turned into the wind, and ran through his checklist: controls free, fuel valve on "both", take-off trim set, mixture full rich, prop set to climb pitch, magnetos checked, flaps set. He started the takeoff roll and realized how distracted he was by Onésimo's injuries and constant shrieking when he felt a breeze. The passenger window was still open – a standard pre-takeoff step he had missed.

Chuck reached across the nurse in the right seat, closed the window, and a little more focused than before, took off for Pucallpa.

JAARS pilots fly according to the highest operational safety standards in bush aviation. Altitude is life above the "broccoli," pilot-talk for the bushy, hundred-foot trees they routinely fly over in the jungle. The flight plan called for six thousand feet cruising altitude back to Pucallpa. But the greater altitude posed another risk to the screaming child in the seat behind him. Chuck climbed to four thousand feet and leveled off to minimize the potential negative effects of altitude on Onésimo's head injuries.

But fifteen minutes into the flight, Chuck's heart sank, and he began to think he should turn around. Onésimo had stopped screaming. Chuck was sure he was dead. But when he looked over his shoulder to check, he could see that the boy was still breathing.

A cycle of waking terror and unconsciousness set in which continued for the rest of the flight. Onésimo would lapse into sleep for a few minutes and then wake up shrieking until his strength was spent. Chuck's nerves were frazzled by the time the Pulcallpa airport filled the windscreen. The weather had been clear, and Pucallpa air traffic control kept the rest of the aircraft in a holding pattern while Chuck made a base leg approach to final.

Upon arrival at Pucallpa, they learned that the ambulance they had called had not waited. Chuck put Onésimo, his mom, and the nurse in a taxi and sent them on their way. Standing there on the ramp watching the taxi leave, Chuck felt a little letdown. This was what he had become a missionary pilot to do. But he still had a hangar to close. He continued to pray for Onésimo as he returned to his work.

Meanwhile, Shuro stood in the hospital waiting room, staring at the door. She had never seen a door before she got on the big noisy thing that flew like a bird. She wanted to get to the other side of the door. She wanted to be with her son. But the men and women in white coats would not let her. So she stared at the door, wishing it to open. She had seen the men and women pulling on the shiny round thing in the middle and to the right on the door. But when she pulled it did not work. There were other men and women in the waiting room, sitting on pads with four stilts under them. Shuro had never seen these things and she did not trust them. She went over to a corner and sat on the floor.

Later that day the phone rang in Merrilee Goins's house in Pucallpa. Merrilee is a Wycliffe Medical Assistant (MA) who worked for a while with Onésimo's Chitonahua tribe. Her supervisor Elisabeth was on the other end. "Can you still speak Chitonahua?" she asked Merrilee.

Shortly thereafter, Merrilee could see and understand Shuro's confusion as the young translator arrived at the hospital. She spoke in Spanish to a sympathetic nurse nearby, "Spend a day and a night in the public buildings of a big city, and count the number of different door and sink handles you use. Imagine that you never saw any of them before and that you are unable to read any of the signs or ask a question in the local language. You're unfamiliar with the money, and had your first ride ever in a four-wheeled vehicle yesterday and your first airplane ride the day before that. Now remember that your little boy has just been mauled by a jaguar and try to figure out what to do next." The nurse nodded her understanding.

Someone has said that the ability to think in another language is like having another soul. Unless you've lived in a different country for a while, it's hard to comprehend the mental gymnastics missionaries perform every day. Merrilee is an American. Missionaries call that her first culture. Peru is in South America with Spanish as its primary language. That's her second culture. The Chitonahua people group is pre-Hispanic Peruvian. That's her third culture. Merrilee was trying to interpret her second culture to a grieving and frightened mother from her third.

Later that night, the reconstructive surgeon came out and told them that Onésimo would have to be flown to the capital, Lima. Shuro was not a Spanish-speaker and was already in culture shock in Pucallpa, in addition to the horror of Onésimo's trauma. How could she possibly cope in the big city?

Elisabeth gave Merrilee permission to go with them and sent out word of Onésimo's condition by email. Many, many people began to pray.

Chuck Daly had been praying all along. Just a few weeks later, as he was making his way back to the States, he stopped in Lima to visit Shuro and Merrilee. Chuck prayed with them and headed home.

Epilogue

By the grace of God, little Onésimo made a full recovery. He still bears the scars and his left eye has no sight, but those who know him well say that he is very cute and brighter than most.

Meanwhile, Chuck Daly returned to the States to begin his work as Aviation Director for JAARS in Waxhaw, North Carolina. If you ask him why he remains in mission aviation, he will tell you it is the joy of seeing lives transformed by God's Word. But the impact of his life's intersection with the little Chitonahua boy remains.

"For a year or more afterward," Chuck says, "I couldn't look at a picture of a jaguar or tell the story of Onésimo without tearing up. Even now it still gets to me."

Chuck is now the JAARS Vice President of Global Transportation Services.

Bangui satellite dish. Where you use the stuff
you learned in math.

7

Spinning the Cosmic Cube

*The longer one lives, the more one realizes that every-
thing depends upon chance, and the harder it is to believe
that this omnipotent factor in human affairs arises simply
from the blind interplay of events. Chance, Fortune, Luck,
Destiny, Fate, Providence, seem to me only different ways of
expressing the same thing, to wit, that a man's own contribution to his life story is continually dominated by an external
superior power.—Winston Churchill*

In the film, *The Pursuit of Happyness,* Chris Gardner pursues Jay
Twistle for days, trying to convince the head of recruiting for
Dean Witter Investments to give him a shot at an interview. Twistle
politely ignores the broke salesman, until the day Gardner finagles
a cab ride with Twistle. In the cab Jay is still ignoring Chris, trying
instead to solve his new Rubik's Cube, when Gardner says, "I can
do it."

"No, you can't. Nobody can."

"Yeah, I can…give it to me." With the short cab ride as his timer,
Gardner spins the cube.

Twistle is at first bemused, then intrigued, and finally fascinated
as Gardner, the clock ticking down, hands him the solved puzzle.
He knows now that this is no normal door-to-door salesman. He is

in the presence of superior ability. Before Twistle leaves the cab, Gardner has his invitation to an interview.

When most people think of the power of God, they think of the obviously miraculous or the utterly terrifying – instant healings, addictions broken in a moment of prayer, tsunamis, tornados, earthquakes, and the like – but how hard is that for God compared to spinning the cosmic Rubik's Cube of time and space, circumstance and human free will, into patterns that accomplish his purposes?

JAARS Information Technology specialist Woody McLendon has a clue.

No Bags at Bangui
Saturday, January 21, 2007, 6:00 A.M. LT, Bangui, Central African Republic (CAR):

"I'm sorry, monsieur, but your luggage was not on the airplane."

"Can you tell me when to expect it? When will the next plane arrive?"

"Next Saturday."

Woody McLendon and Paul Zwierzynski, JAARS technicians, had just arrived for a two-week satellite Internet installation project, and most of their equipment was missing. They were in the Central African Republic (CAR) as part of a joint project between the African Wycliffe organization in CAR, and ICDI, another mission. ICDI had bought two other dishes to install in two other cities, Bouali and Berberati. The previous summer the JAARS team purchased the materials and put them into a container to ship by sea. But a delay coming out of customs prevented a November installation. In the meantime, the materials had been divided up among the three cities, Bouali, Berberati, and Bangui.

To make matters worse, the boxes hadn't been labeled correctly, so some things had been shipped to the wrong places. Two critical pieces of equipment for the satellite setup were missing. Not a great start for the mission in Bangui.

But this is what it looked like from Woody's point of view.

Bangui Journal: "My sense is that God is allowing the situation to look grim on the outside in order to show that he's the one making this work, not us... We're still quite encouraged."

Satellite linkups aren't simple. It isn't like plugging in a radio and aiming the antenna. Doing it with missing parts in borrowed clothes on the backside of Africa is like trying to start a fire in a snowstorm. You'd have to know something about Woody McLendon to understand how he could be so cheerful.

Woody, a little over six feet tall with close cropped dark hair, is a soft-spoken man in his early forties. Some missionaries are energetic and full of zeal. Some are often fatigued, physically and emotionally, running right above empty. Some are in great wrestling matches with God, struggling with where he has them, wondering what to do next. Woody McLendon is at peace. His soul is very still. There is great depth there, the kind that comes only to those who have been in deep places with God.

Even though his background was in software engineering, Woody didn't start his Wycliffe journey as an IT guy. He and Mary – his wife of twenty-one years – began as translators. Like many Wycliffe missionaries, Woody was attracted to translation because it looked like a good way to do missions without having to become a pastor. Mary was already a nurse, Woody had earned a Master's in Software Systems Engineering, and they had their first child before Bible translation became a serious possibility. He'd heard about Wycliffe at InterVarsity's Urbana Conference in 1984. Life took some turns after that, and he was reintroduced to the ministry through the Quest outreach that Wycliffe ran in the early '90s. He and Mary began preparation for Bible translation ministry in 1993. Linguistics and the analytical sciences have a lot in common, and Woody enjoyed the study.

Still, it took three years to reach the field. That's when the challenges they faced caused them to begin wondering if they had heard God correctly. They were going to be asking themselves that question for a long time.

Not Just Surfing

Bangui Journal: "Our colleagues here are very excited about having a functional, stable Internet connection, and we don't want to let them down. There is still A LOT of Bible translation work to do here in the Central African Republic, and this will help them be well connected to consultants [specialists in translation, literacy, linguistics] as well as to the administrative leadership here in Africa and internationally.

"One last exciting point. When we came in on Saturday, the first ever Consultation for the Evangelization of CAR was finishing on the Wycliffe compound. There were representatives from 22 or 23 denominations in CAR together for 3 days to discuss how to reach all the people groups of this country. The church leaders recognize Bible translation as a critical part of effective evangelism, discipleship, and church planting, and see Wycliffe as an important partner. Thus our work here isn't about helping people surf the web faster, but about building the communication and collaboration infrastructure to speed Bible translation. We're privileged to see the pieces unfolding before our very eyes."

Pieces of Dream

Bangui, Central African Republic, wasn't Woody's first African experience. He and Mary, along with six-year-old Katie, four-year-old Ruth-Anne, and infant Suzanne, arrived as translators in Niger, Africa, in 1996 and began the transition to life in Africa. Partnership development and linguistics study had used up 1993-1994. Then it was off to Brussels for a year of French, Niger's official language. The dream was coming together. But translation training is not the same thing as translation in the field. The McLendons began running into problems that no one could have trained them for.

It is said of Satan that he really doesn't have to manufacture trouble for Christians. All he has to do is manipulate their weaknesses and magnify their differences, and the job is more than halfway done. Life for Westerners in a third-world town – missionary or no – has a peculiar way of stripping off carefully developed, culturally supported coping skills. And, in the end, missionaries are just people.

By December of 1997, Woody and Mary were feeling pretty vulnerable. Conflicts with colleagues, misunderstandings, and illness combined to form tiny cracks in their dream.

In 1998 the cracks became fissures, and pieces of their dream began falling away. A ruptured disc put Mary in bed with excruciating pain and eventually sent the McLendons back to the States for treatment.

You don't have to be a missionary to comprehend the discouragement, the disappointment that settles in when you have to quit the field early. Woody and Mary had spent the better part of five years preparing to be translators. Other than translator support – office management, bookkeeping, computer repair – the closest they got was a seven-week language survey mission Woody took to the eastern deserts of Niger before Mary's injury.

Stateside, mourning compounded their discouragement. In 1999, as Mary's back was healing, her father, a healthy octogenarian, underwent hip replacement surgery. Complications led to a stroke and he died. With wounded hearts, Woody and Mary continued preparations to return to Niger as translators. A small consolation was that settlement of the estate enabled them to start building a house in Waxhaw near the JAARS Center. The contractor was finishing their house as they made their way back to Niger and planned to rent it out until they returned on furlough.

They did make it back to Niger. But while in the U.S., Suzanne developed a chronic case of eczema. In the States, this is a nuisance. In sub-Saharan Africa, where small rashes easily morph into major infections, it's a menace. Once back in Niger, the eczema returned with a vengeance. Toddlers can't comprehend "don't scratch," and subtropical climates make itching little girls into human petri dishes for infection. Impetigo set in, forcing a painful choice upon Woody and Mary: stay in Africa and endanger their child or return to the States indefinitely and risk losing the support of churches and individuals who signed on to put them on the field as translators.

They went home and, because they would no longer be on the field as translators, they did, over time, lose some of their financial support.

Woody was spiritually and emotionally stunned, locked up inside until finally, after much agony, he realized that he had experienced so much disappointment, so many unmet expectations, and watched the death of so many dreams that he was afraid to dream, afraid to think of what might come next. It took a while, but in the end he managed a new kind of prayer. "Lord, I'd like to be able to dream again, but I'm afraid to dream. So please let me begin to dream your dreams for me."

Slowly, gingerly, Woody's wounded spirit began to reawaken, and a new possibility for serving the task of Bible translation occurred to him. *I wonder if God is calling me to focus on Information Technology.*

Spinning the Cosmic Cube
Monday, January 23, 2007, Bangui, CAR

Back in CAR, Woody and Paul found the copper grounding rods and wire on a day trip to Bouali. These parts provide critical protection for any electronics in the stormy subtropics. What they didn't find were the satellite modem and the coax cables. No cable means no signal could flow from dish to computer. No modem and, even if the cable was good and the signal strong, no computer could descramble it.

That's about the time that Woody began to see the cosmic Rubik's Cube spin.

They didn't have the modem or the coax, but they did have a satellite phone. And they knew that Jim Hocking (ICDI director) was flying back from Indiana to Cameroon the next day. It was 6:00 A.M. on the East Coast in the U.S. when they reached one of the technicians at the JAARS Center in Waxhaw, North Carolina, and he went to work on the problem. Within a few hours, he had found replacements and had them FedEx'd to Jim in Indiana. Jim boarded a plane the following night. But he was only flying into Cameroon, still a long way from Bangui, CAR. It could take a week to get the modem and cable from there, but Woody and Paul didn't have a week to spare. One other critical aid they needed was Francis, a Cameroonian computer technician who could greatly speed up their work and be a ready source of technical support for the local staff.

He was supposed to arrive by car in four days, but his transportation fell through.

That's when God intervened.

Woody and his friends *just happened* to have dinner with an SIL family in Bangui that Monday night. The husband and father *just happened* to be in the western part of CAR, working with two translation teams. The SIL plane from Cameroon *just happened* to be flying him back to Bangui on Thursday. Jim Hocking's flight to Cameroon *just happened* to be arriving on Wednesday. It *just happened* that the flight had room for Francis to hop a ride directly to Bangui. And one more thing, even though their high tech aiming equipment was stuck in their luggage with Air France, Paul *just happened* to be an astronomy and trigonometry whiz. He used the sun and some fancy math to locate the satellite in the sky and get the initial aiming done for the dish. When Francis arrived in Bangui at noon on Thursday with the modem and coax cable, they quickly finished the configuration and powered up the dish. It was dead on target! None of the missing equipment was needed after all.

Dreaming God's Dreams

At JAARS, the IT team needed help, and Woody was an obvious fit. In 2003, shortly after losing a major portion of their financial support because they were serving in the U.S., Woody was working in a JAARS office in Waxhaw. Unbeknownst to Woody, the Missions Committee chair from the church he attended in college was visiting Waxhaw for another matter. Woody was still in the throes of asking God, "What do we do about support? Whatever it is, we know you have us where you want us, and we trust you." The man *just happened* to drop by Woody's office.

After catching up with each other, Woody asked, "Whom would I contact at the church to ask about financial support?"

"That would be me."

"What do I need to do?"

"Write a letter, explaining how God has been leading you, what your ministry is, and what your needs are."

That church is now a major supporter of the McLendons.

Ministry life is often fraught with disappointment. God some-times allows what seem to be horrendously unfair, sinful, or terri-fying circumstances – even major breakdowns in relationships with fellow believers (see Paul and Barnabas in the book of Acts) – in order to move his people into places they would never have consid-ered going to before. He has to make the current situation completely untenable, almost physically break Christians out of their comfort zones, before they will leave what is known for the unknown, their dreams for his.

Bangui Journal: "It is amazing to believe that only 72 hours after we requested it, we had the equipment in hand in the middle of central Africa. We are all amazed and praising God for providing for us…. This did not happen because of us!"

Epilogue: February 2, 2007

Bangui Journal: "In just a few hours I board the Air France plane that will take me to Paris to begin my journey home. By God's grace we accomplished all that we needed to do….

"Yesterday, my colleagues worked at ICDI and got their tower assembled with the wireless network antenna on top. When they put it up, the wireless network, spanning over 3 miles, worked the first time!

"Today was a very special day. At noon, we had a celebration and dedication time – all the ACATBA (the Wycliffe organization in the Central African Republic) members, the SIL members, many from ICDI, and those of us from JAARS all gathered together to thank God for what he has done and to dedicate this equipment to his service."

Woody McLendon now serves as JAARS Vice President of Global Information Technologies in Waxhaw, North Carolina.

David Reeves and family circa 2002

8

Escape from Island M

❧❧

But whenever they persecute you in this city, flee to the next…
Matthew 10:23a

David Reeves could not yet hear the screams of the dying, but he could smell the smoke and across the bay see flames engulfing the harbor town of fifty thousand souls.

The jihad had begun.

Only one month before, in December of 1998, David had taken over as the Island M Aviation Director (AD) for JAARS. His new job had given him oversight of flight operations, scheduling, ticketing, maintenance, and all the responsibilities these tasks entailed. He was essentially running a small, single ship airline – only on this island, his plane landed on rugged, often bomb-cratered airstrips left over by the Japanese in World War II.

One week prior to the jihad, Summer Institute of Linguistics (SIL) had sent an instructor to the island to do contingency training. Contingency training prepares missionaries to handle the threat of violence or rape or terrorism. "The idea is to be proactive," the teacher explained. "Don't wait to be a victim. Do something before it's too late." The instructor was gone only a few days before large groups of loosely organized jihadists came down from the hills in trucks, leaving behind a wake of destruction that encompassed entire villages, burning houses, raping women, and hacking people to death with machetes.

SIL was hosting a Mother Tongue Translator workshop in the town across the bay from David's house at the time, training about thirty helpers from various villages to be more effective as Bible translators or translation helpers. They had a lot of people in the small city when news of machete murders and Molotov cocktails came down from the hills. The rumors were that Christians had set fire to some mosques, which turned out to be false, but played well in the hands of the foreign-born jihadists inciting the violence.

That same day, in an area not far from the Reeves' home, a village pastor heard trucks rumbling down the road and shouted to his wife, "Get the children. We must run!" He and his wife grabbed their children by the hand and fled to the jungle. Other villagers joined them. The jihadists jumped out of the trucks screaming, coming closer, and firing their weapons in the air. The trail grew dim and diverged. The pastor took his family to the right. The others split off to the left, including a young woman, heavy with child. She scrambled as best she could through the dense undergrowth, but she was no match for the crazed men. She stumbled one last time and the jihadists were upon her, hacking with their machetes, and howling like mad dogs. One man threw her on her back, and as she screamed for mercy another slashed hard – opening her belly. He yanked the child from her womb, killing it before her eyes, threw it on the ground, and turned his machete back on her until she lay lifeless on the jungle floor.

Meanwhile, David and the other missionaries were sticking close to home, unaware of the extent of the chaos, trying to confirm the reports of violence.

Josh Jones, one of the translators helping with the Mother Tongue Translator seminar, voiced his concern for the village folks helping in the workshop. "I feel responsible for the people helping us in the translation work. If I leave now," he said, "I can probably beat the mob and get these folks back to their families."

David knew that the town was on fire, but he couldn't confirm the killings. The mission leadership weighed the risks and eventually agreed that Josh should take the ministry minibus, an oversized

Mitsubishi minivan, and try to get them home. After he was gone, David heard the story of the murdered pregnant woman from the pastor who saw it.

On Tuesday rumors and horrifying reports of killings had emerged from the burning town across the bay and surrounding villages. Josh drove into harm's way on Wednesday, one day into the conflict. By nightfall he had not returned. No one knew his whereabouts, and with cell towers jammed and telephone exchanges burning, they had no way to contact him. By Thursday some of the mission's men had identified the mutilated bodies of national friends, lying on the floor at the police station. The reality of brutal death hit them hard. They now knew that more than fifty Western missionaries were locked in the crosshairs of jihad.

Physical danger is not the only pressure a missionary feels in this kind of situation. The Reeves had only been back in the country for a few months. It had taken over a year of hassling with red tape to obtain the visas. All of the JAARS and SIL workers had rental contracts on houses that had to be paid several years in advance. Added to that were the relationships they had built with nationals, the progress on the translation work, and the safety of the local people who worked for them at their homes, at the hangar, and as translators. David and his colleagues felt *obliged* to stay as long as they could to see if things would settle down. But they could feel the threat rising.

Thursday dawned thick with the stench of the still burning town across the bay. No one had slept much, and soon they were all at David's house trying to figure out what to do. The Contingency Committee, a group of five people, was responsible for *everyone's* safety. A full review of the situation revealed that they had but two options: flee to the military bases with all the rest of the refugees – which by this time numbered in the tens of thousands – with no water, no food, no shelter, and little hope of getting any. Or they could leave. They had made prior arrangements with Airnorth, a small charter airline out of Australia, to evacuate them should the need arise. But with the phones dead, they couldn't contact Airnorth. To make matters worse, they were sick with worry about Josh and regretted ever having let him go.

JAARS was founded for the purpose of providing safe, efficient travel and communications services for SIL translators. It was beginning to look like they'd lost one of them.

Then the familiar clatter of the old minibus broke the tension in David's house. "It's Josh!" someone shouted. "Josh is back!"

Josh Jones was welcomed as one raised from the dead. But he also confirmed their worst fears about the violence.

"The whole town has burned," he said. "I was on the way back from one of the villages when I practically ran into a dump truck filled with jihadists. The telephone exchange was a hundred yards up the mountain. It was in flames. The men in the dump truck were shouting and waving their machetes in triumph. I stopped the minibus, not sure what to do. Then they saw me and started to move. The gate to the military base was on my right. I slammed the bus into gear and drove hard for the gate before the truck could cut me off."

He was safe – for the moment, but outraged that the soldiers were doing nothing to stop the barbarity. Josh took his life in his hands and confronted one of them, "Why aren't you stopping them? Why are you letting them destroy the town and kill these people?"

The soldier just looked at him and said, "We have orders not to intervene."

With the chaos surging all around the base, Josh couldn't risk leaving it. He spent the night within its walls, getting what sleep he could in the minibus.

The room fairly exploded in conversation after Josh's account. The Contingency Team and others present weighed options and risks, means of transportation to the airport, how to get safe passage, how to protect homes and property, whether they should stay or flee, and whether flight was even possible. In the midst of all the discussion, someone said, "Maybe we should go home, sharpen our machetes, and get ready to protect our families."

That comment crystallized something in David's mind. "Do you hear what you're saying?" he asked. "I didn't come here for a war. If we're really talking about kill or be killed, it's time to leave. Let's get what we can carry, contact Australia, and get out of here."

But contacting Australia was easier said than done. And the jihad was moving in their direction.

The meeting broke up, and David went to find the nearest High Frequency radio transmitter. His wife, Jane, sat down with their two children, Naomi age eleven and Josiah age eight, and said, "Kids, we have to leave. I want you to get one change of clothes and one thing that is special to you. We can't take anything else."

Josiah came back with his favorite box of Legos, big enough to fill a small suitcase. "Josiah, I know this is hard," his mom said, "but we can only take twenty kilos (44 lbs.), just one suitcase for all four of us. You can only take one toy."

Josiah had just celebrated his eighth birthday. "What about my other toys, Mom? Will we come back for them?"

"I don't know, Son."

The Reeves' house, on a good day, was a twenty-minute drive through many small hamlets from the JAARS hangar and its high powered High Frequency (HF) radio. Each little village, this one Christian, that one Muslim, was now roadblocked. With that radio inaccessible, David, whose specialty was radios, had only a friend's low powered HF radio to call a neighboring island and ask them to contact Airnorth with news of their plight. On hearing the request, the president of the small airline scrambled two Embraer EMB 110 regional turbo-props into the air as fast as possible, flying one of them himself. But he was flying into a war zone. Things were about to get interesting.

With Airnorth enroute, David and his colleagues began to pull together a plan to get to the airport.

JAARS personnel were able to convince the Army, for a fee, to escort a group of five cars to the airport. Some of the SIL people, stuck in town across the bay, were able to pay the Air Force commander enough to provide one of his busses. They rode for forty-five minutes through the burning town and around the bottom of the bay to meet David's group. Then they convoyed the twenty-minute ride to the airport, the military escort opening the village roadblocks for them along the way.

While David's group was dealing with roadblocks on the ground, Airnorth's president Michael Bridge was facing a different kind of obstacle. It, too, came over HF radio, from his insurer. "If you land that plane in a war zone, we will cancel the insurance coverage for your airline."

Bridge didn't ponder the financial implications for an instant. "Do what you have to," he answered. "I promised these people we'd get them out. We're getting them out."

Airnorth landed at the closed airport on Island M that afternoon. But once on the ground they couldn't depart into the airspace over a war zone without government permission. Without clearance they might be shot down. Permission was not forthcoming.

Airnorth had had the foresight to stock the galleys of the planes with fast-food. They served it to the hungry, stressed-out refugees, who ate it gratefully, not knowing where their next meal would come from. But night was fast approaching. The children were getting antsy. The military had left a few guards around the hangar, and after dinner some of the kids began talking to them, fascinated with their AK-47 assault rifles.

Time slipped by. Darkness settled on the airport. The smell of the burning town wafted down the bay, reminding everyone of the menace close at hand. What if the clearance didn't come? They were already out of food. What if they ran out of water? Then David saw a missionary's young son with an assault rifle in his hands. "No! This will end right now!" he said as he took the gun from the boy and gave it back to the soldier. "We aren't taking any chances of someone getting hurt. I don't care if the safety *is* on." The tension, which had been somewhat relieved by the meal, returned.

With a clearance looking ever more doubtful, the missionaries began improvising sleeping space for the smaller children in the limited room afforded by the hangar. But at last, around 11:00 P.M., a Cabinet Secretary's office in the national government cleared the planes for departure.

They loaded the planes to the bulkheads, so close to maximum gross weight that they were weighing everyone and every piece of baggage before they allowed them to board. One rather portly woman refused to stand on the scale, too embarrassed to hear her weight called out. She relented when she was allowed to hold her luggage while being weighed. The smaller children sat on their parents' laps; some teenagers were three to a row. Fifty-one people squeezed into two planes designed to carry only twenty-one apiece. And then they were airborne.

As the planes banked to the south, the fires of war came into view, bittersweet to the people escaping Island M. They were safe. But their friends below, national translators, airport staff, house helpers, shopkeepers, and neighbors, Muslim and Christian alike, were surrounded by death. With hearts heavy for all who were left behind, they held their children close and prayed.

Epilogue
March 1999, monsoon season in Darwin, Australia

Rain drummed hard and constant on the camper roof above his head. All of his equipment, the sophisticated tools of an avionics technician, was gone, and David, for the first time in a long time, found himself with little to do but pray and think.

"I didn't sign up for this, Lord!" he prayed. "I didn't sign up to work in a war zone and see my friends butchered and escape with just the clothes on my back. I think I'm ready to go home and flip burgers for a while."

But the longer he prayed and the deeper he dove into Scripture, the more he heard the gentle voice of his Master saying, "Yes, you did. It *is* the norm for my servants to suffer on my behalf, and some to die. You simply grew up in a place that wasn't normal."

The Reeves lived for five months in that camper in Australia and lost everything they'd left on Island M.

Some families stayed in campers for a year. Many of the missionaries helped national friends escape war torn Island M by paying for boat tickets out of their own meager funds. But often they could not locate their friends, and many they knew were among the ten thousand souls lost to the jihad.

Late in 1999, David and Jane, Naomi and Josiah took up a different assignment in Southeast Asia, where David served for seven years as a YAJASI team member before returning to the States. It was on that island that he began to see in a new way the part he played in God's larger plan.

David Reeves camps next to downed plane.

9

Single Rope Technique

"And who knows but that you have come to royal position for such a time as this?" Esther 4:14

"How'd I ever get myself into this?!" David Reeves muttered into the rotor blast of the Helimission Bell LongRanger. Strapped into the back seat, inches from an open door, David and his ministry partner, Mission Aviation Fellowship (MAF) pilot Dennis Bergstrazer, felt the single turbine spin up to full power. Smoke from the breakfast fires of Nabire, Papua, Indonesia (formerly Irian Jaya), mingled with fetid jungle and penetrated his nostrils until the helicopter found its wings and lifted them beyond the morning mists.

He and Dennis were on a rescue mission. A plane was down in the mountains, a single engine Cessna 185 operated by another agency. It had fallen into a ravine while on a timber surveying mission four thousand feet above sea level. No roads or trails led to the place. The jungle was too thick and the ravine too steep for a chopper landing. The only way in, the only way to reach any possible survivors, was to drop from a helicopter using "single rope technique."

Single rope technique, or SRT, is the skill of mountaineers, rescue workers, soldiers, and that truly odd lot of explorers called cavers. Mountaineers do it for the adventure. Rescue workers do it to save lives, and soldiers do it soldiering. But cavers' motives, well, cavers are a different lot. It's dark down there, an inky, black, oppressive kind of dark. There are pits as deep as forty-story buildings leading to little holes that men weren't meant to squiggle through. Cavers make a game of rappelling into the pits, finding the holes, and squiggling through them. For this they use SRT. David Reeves was a caver way back when. Good thing too, as SRT is not a child's game. The equipment is purpose built, the rope specially made. Run the rope through the rigging exactly right, and it's more fun than should be legal. Run it wrong and, well, they don't call it a "death rig" for nothing. There may be a few hundred non-military people on earth who really know what they're doing with SRT equipment. David Reeves is one of them.

JAARS folks don't have flashy exteriors, and David is no exception. He is a mild-mannered avionics technician and aviation program manager. He wears conservative clothes and drives a minivan. He is tall and middle aged. His speech is still influenced by his western Georgia upbringing and, while not monotone, is not what one would call expressive or dynamic. But that easygoing exterior belies the adventurer within. As a caver in his twenties, he regularly dropped six hundred feet into pitch black caverns on naught but a single rope. He served on rescue squads. He's traveled the globe many times, negotiated with headhunters, cannibals, developing world governments, and heads of large corporations. He's led mission expeditions into the hinterlands of the uttermost parts of the earth.

David is also a technological innovator and entrepreneur. He and JAARS colleagues Russ Perry and Carman Frith developed the Automated Flight Following System (AFFS) now in service in most JAARS aircraft and also used by some mission and other aviation programs that operate in remote areas. He also led the effort to make YAJASI, the JAARS partner in Papua, a dealer for the revolutionary satellite phone. David is an adventurer, but a sober one. He thinks things through.

That's what he was doing when his fellow rescuers began to question if they should go at all. "We know *where* it is," said Wally Wiley, Director of Operations for MAF which was aiding the effort, "but we can't make contact. No one knows if they're dead or alive. I'm not sure we should risk anyone else if there's no one there to rescue."

The head of the agency that owned the crashed plane added, "There's a big winch-equipped military helicopter on its way from Sorong, and the Helimission guys who went up early this morning couldn't rouse any survivors with their horn."

"Has anyone nailed down the location of that military chopper?" David asked.

"No, not really."

"Until we know where the big chopper is, we don't know if it will get here in time. And until someone is on the ground at the site, then no one knows if there are survivors. We should keep moving in that direction until we know for sure one way or the other."

Even David couldn't believe he was hearing himself say this. Just a few hours before, he was questioning his own participation in this high risk search and rescue (SAR). The situation at home was too unstable.

Sentani, the town that hosted JAARS / SIL operations in this part of Indonesia, had been burning for two days. Social unrest that had devolved into riots had everything shut down or severely curtailed; the schools had been closed. Westerners moved about only by necessity.

David, his son Josiah, and many of their colleagues were across town playing softball when the news came that rioting had begun. After waiting a few hours, they pooled their vehicles and caravanned back across town. Everything had been locked down since.

David's wife, Jane, relived the moment in an email to friends the following week: "I started pulling things off the walls and loading trunks. We could see smoke from downtown and hear gunshots. The power was off for a while. The main pharmacy, hardware store, and other little shops with homes above were burning. Further down the street, several more stores were burned down, including the photo developing store as well as a photocopying store that was owned by some Indonesian friends. It felt like the buildup to the evacuation from Island M all over again."

"You have to go, David." JAARS pilot Paul Westlund knew what David was wrestling with. Sentani was in flames, and the military was moving in. Everyone felt the déjà vu.

"Paul, I may need to evacuate my family in twenty-four hours. How can I go off and leave them in the middle of this?"

"You're the only one who's qualified on SRT, David. I'd love to go, but I don't know what I'm doing and you do." This was true. Westlund is an avid adventurer and extreme sports enthusiast. He would have relished the drop into the jungle to save a life. "Remember Esther?" he said. "For such a time as this? You're the man."

So at 5:30 Tuesday morning, an anxious Jane climbed into the Reeves' old Kijang utility vehicle, cranked it up, and backed it out of the driveway. The stench from burned-out buildings punctured the tepid air as she and David eased across the main road toward the airport. The smell was almost overwhelming, but it was the only thing that met them. The town was otherwise quiet. She kissed her husband good-bye, put the car in gear, and slowly drove back to the house to wait.

These thoughts were still in David's head as the Helimission chopper topped a ridge and came to a hover. It is the most difficult maneuver in the helicopter's repertoire, the hover. Full power on the collective, stick and rudder pedals constantly moving in tiny increments, the pilot balances his machine at the pinnacle of its performance envelope and holds it there. Sometimes all it takes is a gust of wind to push it off the point. But the Helimission guys are good at what they do. At forty-five hundred feet above sea level, the LongRanger kept her poise.

Five hundred feet below them, David could see the white plane with brown and gold stripes, over on its back, nose crushed, a single gear leg sticking up in the air like a white steel arm reaching for help. He could see the point of impact, the tail cone snapped off like a twig, hanging by its cables. And he could see something blue out on the wing. The chopper began a slow descent into the ravine. At three hundred feet David yelled into his headset.

"Dennis, can you see what's on the wing?"

"That blue thing?"

"Yeah."

"What is it?"

"It's a man!"

They tried to signal the man, but he made no reply. Still, he was sitting up, and they knew what they had to do. Brian Smith, the Helimission pilot, nudged the stick forward just a tick. The LongRanger tilted itself out of hover and sailed down the ravine. Smith's team had already cleared a landing zone (LZ) on a sandbar four miles downstream. They would rig the rope there.

"Dennis, you have the most recent medical training so you should go in first." David was checking the French-made Petzl rappelling gear Dennis was using. Smith was removing the roof panel below the rotor. This Helimission LongRanger wasn't built for SAR or SRT. There was no place to anchor the rope except on one of the engine mounts that could be accessed behind the panel. As soon as he was sure Dennis's rig was right, David tied the knot on the engine mount and backed away. The chopper wasn't powerful enough to hold a hover and drop Dennis with David on board.

"Get down as fast as you can," he had told Dennis. "If this thing comes out of hover and starts to move, you're a goner. The miners lost one of their geologists like that a while back. Chopper flew him into a tree when he was still on the rope."

David shielded his eyes as the chopper cleared the sandbar. It wasn't until then that he realized he was standing in the middle of a river that was a two-week jungle hike from anywhere civilized. If the chopper didn't come back, he was stranded.

As they rose from the sandbar, Dennis and the pilot realized they didn't have much time. Besides fuel limits, they had to think about mountain wind currents as the day wore on. The equatorial sun begins to boil the jungle early. Thermals, rollers, and wind shears become so unpredictable that JAARS operates with wind curfews on many mountain strips, some of them beginning as early as ten in the morning. This was no airstrip, but a steep gulch thick with trees, and no margin for error. They decided to drop Dennis a little below the crash site, no more than twenty yards away.

The chopper had to fly back to Nabire to refuel then back to the sandbar LZ to pick up David. About an hour after they dropped Dennis, David was hooked in and ready to go. Brian, the chopper pilot, twisted the throttle, brought the engine back up to take-off power, and as he lifted the collective and rose from the sandbar, spoke to David through the intercom. "Dennis took a long time to get down. I don't want to do that again over the trees. I spotted a site a little above the wreck that looks like a landslide. Can you go in there?"

"Yeah!" David shouted back. "And the gear I'm working with is faster than the Petzl. I'll get down in a hurry." Soon they were over the site. Brian spun the Allison turbine up to full song and brought his bird into a hover. With a quick prayer, David was over the sill and squatting, back to the wind, on the landing skid. He held the rope with his left hand in front and his right hand behind. Normally, at this point, he would tilt back till he was almost perpendicular to the ledge and kick out to begin the descent. But David is a big guy, and the LongRanger is not a large aircraft. He didn't want to offset the center of gravity and throw them out of hover so he shinnied himself down off the skid. Once his head was clear, he let the rope slip.

Rappelling down a mountain face is thrilling. It's also pretty predictable. The rope is anchored to something solid above you. Every now and then, your feet touch the face and you push away or stop a twirl. And if you've done your homework, you land long before the rope runs out.

Rappelling from a hovering helicopter makes mountain rappelling feel like the playground at Burger King. The rotor wash whipped at his clothes and spun David around. The ground moved up and down as the pilot worked to hold the chopper steady. The line zipped through the carabiners as David rushed down one hundred twenty feet. Ten feet above the ground he hit the brakes, slowed till his feet took his weight, and, slipping and fighting for footing in deadfall and loose dirt, unhooked his rig. He was only a few yards from the wrecked plane and the man he was trying to save. He waved to the Helimission crew as an unmistakable odor wafted across his face. *Avgas!* He thought. *This thing could light off in a heartbeat.*

Noise was the other thing that met David in the jungle ravine that Tuesday morning. The Emergency Locator Transmitter (ELT) was still signaling, screeching its alert across the channel on David's handheld aviation radio. "David, can you give us a read on the survivor? What's his condition? And can you cut that thing off?"

Brian asked from the chopper above. The whup-whup of its rotors bounced out a noisy racket down the ravine.

"Can do. Give me a minute!" He was stumbling his way down across the uprooted trees and rocks in the landslide thinking, *Where in the world is Dennis?* But Dennis was still hacking through jungle.

David could see the Cessna more clearly now, its nose flattened, instrument panel tossed out the windshield like so much baggage. The engine had been driven back into the cockpit. People in the front seats had received massive injuries. And no one was calling for help. He moved closer, not wishing to see dead men, but unwilling to give up hope. He saw fuel running down the upper side of the downhill wing, dripping into a puddle of leaves and grime. "There's the leak," he said to no one in particular. Then he heard a groan. Pak Chandra was sitting where they'd first seen him, out on the wing, his feet in the stream.

David had just seen Chandra when Dennis came hacking and sweating out of the jungle. David's first thought was *Where have you been?* But then he saw how thick the jungle was up there. Dennis had been working hard for over an hour to get to the plane.

"Dennis, would you mind checking the men in the cockpit?" he said.

Five men had been in the Cessna 185 when it took off the day before. Pak Chandra was the only survivor. David pulled the man out of the stream and dragged him up onto the uphill wing. Dennis emerged from his inspection of the crumpled cockpit and shook his head. "It looks like one of them survived for a while. But they are all gone now. Let's see if we can get to the ELT and shut it down."

This was no simple task. The Emergency Location Transmitter is located in the back of the plane, behind the seats, near the battery. One spark could ignite the fumes of the hundred octane avgas pooling beneath them. They worked together, prying open the access panel, using pocket multi-tools to disconnect the battery and disengage the transmitter. With the radio frequency quiet, they could hail the Helimission chopper.

"The patient is in a lot of pain and he's in shock. He can barely talk. But his pulse is good. No other survivors, Brian," David spoke into his radio.

"OK, we're out of here for fuel and supplies. We'll be back ASAP."

"Roger that."

Chandra took a little water, a few crumbs of food from his rescuer's first-aid pack, and some aspirin. Then David covered him with a blanket from his kit and moved around the aircraft to see what he could do about the leaking avgas. He and Dennis fashioned one of the fiberglass wingtips into a small gutter and laid it under the dripping fuel. This allowed the pungent avgas to run down into the stream to be carried away by the water instead of pooling beside the plane as a fire and breathing hazard.

"I wish we could talk to a doctor about this guy," Dennis said worriedly. "I feel way out of my depth here."

"Not a problem." David opened his pack and pulled out something that looked like a big cell phone with a fat, cigar-like antenna.

"What are you doing?"

"Satellite phone," David said with a smile. "Let's call the doctor."

The big chopper was still a day away. Having given Pak Chandra the best care the satellite could summon, they set up camp for the night. Dennis took the part of the wing between their patient and the fuselage. Chandra they moved further out, closer to the wingtip. This was partly so they could get to him easier, and partly to shield him from the bodies of his friends still hanging upside down in the plane. David hunkered down just forward of the wing beside the cockpit.

He was straddling one big rock. A congregation of smaller ones dug into his hips, while the wing root supported his back, right underneath the shoulder blades. Four lifeless men hung upside down a couple of feet from his right elbow, still strapped into the mangled plane, death now blending its scent to that of the avgas. It was not a setting conducive to sleeping. Just before sunset, something familiar caught David's eye. Lying just forward of the instrument panel, which the impact of the crash had tossed toy-like out of the windshield, he saw a battery half buried in the dirt.

Dennis heard him rooting in the earth. "What is it?"

"It's a video camera battery," he said as he continued to sift through debris. "And there's the camera! Maybe that will tell us what happened."

David stowed the camera in his bag. It would reveal that the pilot had made the simple, if classic, mistake of many mountain aviation crash victims, flying into rising terrain where the altitude exceeded the performance envelope of the fully loaded airplane.

Dennis smiled down at him from his perch on the wing. "Keep your eyes peeled down there, Buddy."

David chuckled. He knew what Dennis was kidding him about. Papua is not known for exotic or deadly animals, except for the Death Adder, a short fat snake that kills with one bite.

"Thanks for reminding me! You're up on a wing, and I'm sleeping on the ground with the rocks and the critters!"

Darkness comes quickly on the equator, and faster yet in a ravine. Thousands of little eyes reflected the battery powered camp light as David and Dennis took turns watching Chandra through the night. It was creepy, but thankfully, no snakes appeared. Every now and then they adjusted the patient to keep him from rolling off the wing. It was a long night for the rescuers, and a painful one for their new friend. But it was one he would survive.

David lay still as first light crept into the ravine. In a few minutes he would switch on his portable aviation radio to listen for the incoming chopper. Chandra was resting quietly. David could hear Dennis snoring softly. Soon he would have to wake them, break camp, and make ready to be hoisted out of the crash site. But for now his mind wandered and wondered at the providence behind it all.

In 1983 David taught Wally Wiley the Single Rope Technique, just for fun, in Woodstock, Georgia. Wally was now the head of Mission Aviation Fellowship, another organization working in Papua. Wally had taught Dennis what David had taught him. And now, here, eighteen years later and halfway around the world, God had used those skills to save a man's life. He felt privileged just to be there.

A couple of years back, David had been ready to leave the field, the tragedy of Island M having been almost too much to bear. The recent riot had raised those thoughts again. But now he wondered, *What would I miss if I did that?*

He reached over and turned the knob on the radio.

Epilogue

The Reeves served for seven years in Papua, Indonesia, before returning to the States to serve JAARS in an administrative capacity. David became the eighth president of JAARS on October 1, 2008.

A Brief History of JAARS

Through partnerships worldwide, JAARS provides quality technical support services and resources to speed Bible translation for all people.

JAARS used to stand for Jungle Aviation & Radio Service. Its mission has expanded such that the full name was dropped in favor of the acronym alone. It was founded by the same brilliant innovator and evangelical entrepreneur who founded SIL and Wycliffe Bible Translators.

William Cameron Townsend was a Bible salesman who, at 21 years of age, arrived in Guatemala on a mission to sell Spanish Bibles. But not everyone spoke Spanish in Guatemala. In fact over half of Guatemalans spoke Indian tribal languages instead of Spanish. "If your God is so great, why can't He speak my language?" asked an old Cakchiquel man.

Townsend finished the Cakchiquel New Testament in 1929. But by then his vision was much larger than one translation, and he knew that it would take many people to fulfill it. In 1934 he opened a linguistics training school in Arkansas and named it "Camp Wycliffe" after the fourteenth century scholar John Wycliffe, who was the first to get the Latin Vulgate Bible translated into English. The school's name was later changed to Summer Institute of Linguistics and is now known as SIL International. "Uncle Cam," as he came to be known, began to spread his vision of taking God's Word to the Bibleless peoples of the world in their mother tongues.

Lindbergh crossed the Atlantic in 1927. But Cam Townsend was thinking of the potential of the airplane and the wireless to serve mission areas in 1926. After WWII his musings became reality. SIL purchased a government surplus Grumman Duck to fly translators up and down the Amazon River.

But 1947 was the real genesis of JAARS. That's the year that Cam and his wife Elaine were injured in the crash of a small commercial plane in Chiapas, Mexico. Townsend committed himself to developing an organization that could train and equip jungle aviators to serve the needs of translators in speed and safety. The "Jungle Aviation and Radio Service" came to life in Peru in 1948 as a department of SIL. It was officially incorporated in 1963.

JAARS' mission has expanded since the early days. It now has eight departments operating in four major areas: Transportation, Technology, Vernacular Media, and Logistics. In light of this expansion the acronym JAARS was replaced by the word JAARS. The organization is based in Waxhaw, North Carolina.

Cam Townsend died on April 23, 1982, at age 85. He was buried at the JAARS Center two years before David Reeves joined the organization.

End Notes

❧ ❧

This collection of stories is about the people of JAARS. JAARS operates a fleet of specially equipped aircraft in Africa, the Americas, Asia, and the Pacific in service of Bible translation. The fleet logs approximately 1.5 million statute miles per year in about 11,000 hours of flight time. There are about 2400 language groups which do not yet have the Scriptures in their mother tongue. Vision 2025's goal is to get a translation started in each language before the end of 2025.

The JAARS story is inextricably linked with that of Summer Institute of Linguistics (SIL) and Wycliffe Bible Translators (WBT) which deserve books of their own. But this book is about JAARS and the role it plays in the work of Bible translation around the world.

I've spent the last few years, as other duties would allow, observing their work and interviewing people who serve with JAARS. They have all been very gracious and vulnerable, allowing me to see and giving me permission to tell the "gritty reality" of life on the mission field.

The thing we keep forgetting about real adventures, as opposed to the manufactured kind we see in the movies, is that real adventures are uncomfortable, grueling, and often dangerous affairs. The stress doesn't end when the scene shifts. But something kept coming to the fore as I conducted the interviews, something I wasn't expecting. It came when I asked my final question: "You're a competent professional, and this life is hard, even deadly at times. You could make a comfortable living in the secular world. What's keeping you here?"

May God use their answers in your life the way he has in mine.

Note: Please excuse the mystery about some of the exact locations and names of people. JAARS works in some sensitive areas of the world, and the work might be jeopardized if I were to be more specific. If you would like to know more about JAARS, please visit their website at, www.jaars.org.

Thanks for buying this book. It is my hope and prayer that God will use it to help speed the Word to those still waiting.

Acknowledgments

T his collection of stories would have been impossible were it not for my twenty-year friendship with David Reeves and the growing number of friends who work with JAARS. To all who allowed access to their lives and work, patiently submitted to the interview process, and reviewed and checked me for accuracy, THANK YOU for sharing your stories. My life has been immeasurably enriched by them.

My dear friends Scott Foran and Ellen Gray Hogan volunteered their time and expertise to edit the manuscripts and correct my grammar and punctuation. My editor and mentor Sandra Byrd provided a final, professional review of the manuscript, and new friend Jan Harthan polished the manuscript. Even with all this help, errors are inevitable. Any mistakes in fact, grammar, photo credits, or spelling of names and places are my own.

My wife Krista and my daughters patiently endured my absences along with all the other "birth-pangs" of a new author. I love you girls! Friends and members of Faith Community Church in South Boston, Virginia, where I am privileged to be called pastor, also supported the work. Without their encouragement, the book would not have been completed.

Many thanks go to those at the JAARS Center who aided the project: former JAARS President Jim Akovenko, who gave his blessing to start the book; Bud Speck, JAARS Senior Vice President of Marketing and Communications, who reviewed the manuscript for accuracy and compliance; Anna Marie Peterson and her co-workers in JAARS housing who made my visits to the JAARS Center most comfortable; and many others who made me feel welcome.

Finally, a special thank-you goes to my friend who sponsored the trip to Southeast Asia that started it all. Were it not for your generosity, the spark might not have been lit.

There are many, many JAARS stories to tell. I'm filled with gratitude to God that I've been able to share the ones in this book. With his blessing, I will tell more.

Dane Skelton
South Boston, Virginia
May 20, 2009